Melanie Meldgaard

GUIDES AND TIPS

Baldur's Gate 3 is a large and complex game, which offers great freedom of action, both during dialogue and combat. It's not like the majority of modern games, which have a simple gameplay loop. However, here are some tips and best practices that should help you progress more easily.

Character Creation

● You can play an Origin character to discover a particular story, but creating a new character is just as valid. You cannot change their appearance or their race, but their characteristics and even their class can be changed.

● With 11 races, 12 classes, 46 subclasses, and tons of options, it's difficult to give specific advice. You can choose what you want. But if you don't have inspiration, choosing a class with a high Charisma score is a good idea for the protagonist, so you can persuade your allies to act as you wish. Playing a Sorcerer, Bard or Occultist is probably a good idea. If you are not afraid of some spoilers, consult the list of companions, in order to select a different class for example.

● Learn about the right combination of Race and Class for optimal performance. Also remember that you will be playing in a group, there is no need to cover all angles with your protagonist. The most suitable teammate will automatically call on their skills during dialogues and interactions.

● Consider changing your character's name, many players forget this, and they end up with a hero named "Tav".

Group composition

You don't have many options at the start of the game, but having a balanced and versatile group is fundamental:

● You need a class like the Thief or the Bard, which allows you to infiltrate, pick locks, disarm traps, and steal items among other things.

● A martial class like the warrior covers Athletics and Strength checks is also welcome.

● In addition to his many spells, including the fantastic Assist, the priest heals the group, and can cover checks like Perception, Deduction and Religion.

● Unfortunately, the Mage is the only Intelligence-based class, with quite a few associated skills. This feature is also very useful in certain circumstances. It is not obligatory to take Gayle on board, nor to play Magician, but it will make

13

your life much easier.

- Finally, a highly Charismatic character who will handle the dialogues, he must be able to Persuade and Intimidate others is important, at least, if you do not intend to kill everyone you come across. It's even better if it's your protagonist, in order to manage the many situations during which you cannot call on the skills of your teammates.

Exploration

- The world is full of secrets, puzzles and traps. Remember to save regularly, so you can make up for a missed important dice roll at a critical moment. This also allows you to test different approaches during dialogues or fights.

- Also try to visit the area before approaching a critical point, or a big fight. There are often important elements to unlock, which can change a situation. Getting experience and levels is also fundamental.

- Do not hesitate to regularly use the Long Rest at the camp, this allows you to advance the story and the dialogues of your companions. It's almost mandatory to develop relationships between them, and to have access to certain specific quests.

- Many chests and secrets are hidden in hard-to-reach places. You will have to use jumps and even spells to reach them. The Mist Step spell, flight, mage hand, or quite simply, doubling the jump distance, are all ways to achieve them.

- You can view interactable objects with "Left Alt" on PC.

- Inventory management is quite a pain, to reduce the mess in your bags, use the pouches and bags found in the game, then fill them with a specific type of item, such as potions and scrolls. This will help you see things more clearly. Also consider sending food and items that you don't really need directly to camp by selecting the option.

Combat

- When you know that a fight awaits you, do not hesitate to camouflage the members of your group, and separate them in order to place them strategically If your archer is already on top of a roof at the start of the fight, he will be difficult to reach, and he will have a large advantage in attacking the enemy. A thief can also start the fight with a sneak attack from the back, or with an arrow. A spellcaster is also ideal for starting combat, casting a devastating spell to open hostilities.

- Remember to distribute consumables, such as healing potions and scrolls, among the different members of the group.

- The placement of your characters is important, try to quickly place your strong classes near the enemies, while trying to block the passage. They will be forced to stay in melee, or suffer attacks of opportunity to attack your spellcasters and archers.

- Do not hesitate to use short rests, or even long rests, regularly to restore your spells and skills.

- You can also use terrain, such as traps, dangerous surfaces and explosive barrels, to decimate the enemy.

If you encounter difficulties, do not hesitate to switch to Story mode, this will not make the fights trivial either. This will especially give you a boost on dice rolls.

6 MISTAKES TO AVOID TO MAKE GOOD PROGRESS

Baldur's Gate 3 is an incredibly rich game, and you're not going to run out of opportunities to die, along with the rest of the group. Hundreds of choices also await you throughout the adventure with too many ramifications to list everything. However, here are some of the most common errors, on aspects that are not always very intuitive of the game. We limit the spoilers to the first hours of the campaign, in connection with the very beginning of the plot, and what could have been seen on early access.

Neglecting non-combat skills

- Perception could well be the most important skill in the game, since it allows you to unlock additional options during dialogues and interactions. It also allows you to discover many camouflaged objects in the world. Don't hesitate to save often, and load the game if all your characters have failed their Perception roll.

- In the same vein, Persuasion and Intimidation will open many doors, even if their use is more obvious.

- Having a high score in Intelligence will also completely change the situation, in a lot of situations. Unfortunately, unless you are playing a Magician, your character has little chance of having a high score in this characteristic. Killing the group of ogre mages in the first village, in order to steal the crown giving 17 intelligence to its wearer, then equipping it on your protagonist, will make a big difference.

Absolutely try to solve the tadpole problem

Your companions, led by Lae'Zel, will give you the impression that you only have a few days before you start to resemble Doctor Zoidberg. If you take these alarmist warnings literally, it can be tempting to adopt extreme solutions in an attempt to resolve the situation: by forcefully extracting the tadpole. As you might expect, it's not as easy as that, since they tie into the main story. Forget Volo's or Auntie's suggestions, they're really not worth it. On the contrary, they will inflict permanent penalties on you.

Using the power of the Illithid tadpole lightly

Without spoiling too much, once the introduction/tutorial aboard the ship has passed, you will often have the opportunity to use the power of the tadpole present in your later life. The circumstances are varied, it could be to intimidate a goblin, communicate with another tadpole, or even to read the minds of your allies. You will even have the opportunity to devour other tadpoles. Know that this has long-term repercussions.

Depending on the type of character you want to play, and the desired end, you can choose to abuse it, or on the contrary, never touch it. Let's also mention that many companions will be terribly angry if you spy on their thoughts, which can damage your relationship or even block certain outcomes. Conversely, it can also be used to persuade and then manipulate them, but you must not get caught. Remember to save beforehand. The good news is that if you decide to nurture your inner monster supernatural powers can be developed, with a dedicated talent tree and some very powerful powers.

Not getting enough sleep

If you play on a low difficulty mode, or are good at combat, you may not have to return to camp often to take a long rest. But this is a mistake, since quite a few important events in the main story, side quests and companions are triggered during the night spent resting. As said above, you do not have a too severe limit that prevents you from resting, otherwise you will die. Forgetting to sleep is not only bad for your health, but also for your general progress. In summary, remember to take a long rest regularly, especially after discovering an important location, recruiting someone, or performing a special action. This costs supplies, but if you don't overdo it, it's normally not a problem.

Fight honorably

Even Paladins use tactics in combat, you aren't forced to just rush towards the enemy in a group, you aren't playing an orc (a half-orc in the worst case). The deadliest fights can become incredibly easy, if you know what to expect and make the right preparations. In addition to having the right spells, consumables and items equipped to counter the enemy, taking a position before starting combat will be a game changer. By camouflaging your characters, or even using invisibility, it is possible to position the group's archer and mage high up, in a position impossible to reach in melee for example. You can separate character portraits from the group, so you can move your characters individually.

This allows them to bombard the enemy without fearing anything in some cases, such as with the Owlbear. Start the fight with a fearsome spell like Fireball, or a sneak attack. By combining this with spells like grease, explosive barrels placed by you, and others, your melee characters will only have to finish off the survivors.

Neglecting consumables

You will collect a lot of items in Baldur's Gate 3, from food, potions, scrolls, wands and other exotic items. Don't be like those players who kill the final boss with 500 potions in stock. Not only will these items weigh you down, but you may also really need them to win the difficult fights that await you without having to load the game 10 times. Distribute healing potions and the rest of your consumables throughout the group, and use them fairly generously when the situation escalates. It costs actions to open the inventory and transfer, or equip them to a quick slot, you can't easily do it in the middle of a desperate fight.

WHICH ONE TO CHOOSE TO FULLY ENJOY THE GAME?

The difficulty of a game like Baldur's Gate 3 depends greatly on your knowledge of the mechanics of Dungeons & Dragons 5, how you approach each situation, and most importantly, the order in which you approach them. Rushing straight into the final fight of the first act is possible, but unless you know exactly what you're doing, it's probably outright suicide.

Larian Studios has taken these elements into account, and Baldur's Gate 3 should satisfy different player profiles, with its difficulty modes which we will detail. But even in the easiest mode, it is not a "casual" game that will offer you victory on a board in all circumstances.

Explorer - Story Mode

If rules intricacies and tactical encounters aren't your cup of tea, and you don't like to fail your dice rolls, then Explorer mode, aka Easy, is probably for you. It's a bit like your game master is nice and caring, and he lets his players make mistakes without really punishing them.

- Allied NPCs are more resistant, which avoids losing important characters during combat.

- Prices for goods are reduced by 20% from sellers, making it easier to acquire good equipment.

- You have a mastery bonus of +2, this doesn't necessarily mean anything to you if you don't know the game well. But concretely, this will significantly help you succeed in all your attack rolls (hits, arrows, spells) , and skill rolls (persuasion stealth, lockpicking, etc.) and your saving throws so you don't get paralyzed, controlled, and other painful things.

Without completely burdening you, or offering you automatic victories, the story mode will make your adventure more enjoyable. You will have more room to maneuver during big fights, and above all, you will not have to load a save as often after the death of an ally, or a failed roll during a dialogue. This probably won't let you kill a Dragon at level 2, but it's still pretty good.

Balanced - Normal Mode

This is the default difficulty, the one present in early access, and which is supposed to respect the normal rules of Dungeons & Dragons 5. Depending on your way of playing and the situations, a fight can be easy, or on the contrary , impossible. You will therefore have to be careful, and know when to retreat, to come back later. This game mode is also appropriate if you are ready to accept failed die rolls during dialogues, in order to follow the normal course of the situation, without cheating with the saves (even if nothing prevents you from doing so). We advise you to play in this default mode, especially during your first game.

Tactician - Hard Mode

For those who have played Divinity Original Sin 1 and 2, this is the equivalent of Tactician mode. The enemy AI is much trickier, and they will all focus on your most fragile or vulnerable characters. Suffice to say that your Magician risks dying very often. Enemies will also not hesitate to finish off your characters on the ground at the first opportunity, which is much more problematic than a simple fainting.

In addition, enemies will have reinforcements during certain fights, as well as consumables or additional spells. In summary, this will not transform the goblins into super-powerful ogres, but they will be much better prepared, and much more aggressive. It's a bit as if your game master was someone cruel, and who hates you, his objective is then to annihilate your group, while respecting the established rules There's no need to cheat to hurt yourself.

This difficulty mode is recommended from the second part, or if you really want to encounter a challenge. It is better to have in-depth knowledge of the game mechanics, and good, well-optimized characters.

Is it possible to roll the dice when creating your main character?

The answer is no. The old titles of the license let you roll the dice when creating your character, until you obtained a satisfactory score. Some love it, others hate it, since it's possible to waste a lot of time doing it, in order to get ridiculously high specs. In any case, it was an excellent way to modulate the difficulty. If your protagonist has 18 points in almost every attribute, skill rolls and combat are significantly easier. But this is not possible in Baldur's Gate 3, you have a limited number of points to distribute. You can't make the game trivial like this.

From Karmics

This exotic option present in the "gameplay" tab is supposed to smooth luck. If you miss several attacks in a row, then the game will rig the dice, helping you hit your target for example. Or after a critical hit, the next attack has a high chance of missing. This can be a big help in more unbalanced fights. Be aware, however, that this is a double-edged sword, since it also applies to enemies. It has been mathematically proven that you will take much more damage on average from enemies if your characters have high armor. Your tank in particular may suffer. Only use this option if you actually know what you are doing. This tends to make the game easier initially, and more difficult at high levels.

CHARACTER CREATION, RACE, CLASS, CHARACTERISTICS, ORIGINS

Baldur's Gate 3 early access is installed, you have seen (or passed) the introductory cinematic, the time has come to define from what angle you will attack your adventure by creating your character. The process is not particularly long or complicated in Dungeons & Dragons 5 and Baldur's Gate 3, but knowing what you are doing will change things and allow you to be much more formidable during combat or dialogue.

We'll go through all of this in order, you can jump straight to the page you're interested in via the bottom menu.

Custom Character vs. original characters

As in Divinity Original Sin 2, Baldur's Gate 3 offers you a handful of pre-generated characters with very specific appearances, histories and powers. We don't yet know to what extent it will be possible to modify them, but one thing is certain, choosing one of these characters gives access to unique elements. For example, Astarion is a vampire spawn, so he has additional bonuses and penalties, and he seeks to free himself from the vampire who controls him. For the moment, these characters are not available on early access, but they can still be recruited as adventure companions, which also allows you to explore their story.

Custom characters give you greater freedom at every level. However, it remains to be discovered whether their story will be as rich as that of the original characters.

Baldur's Gate 3 : Genre

Drow aside, the so-called civilized society of Faerûn is not sexist. The choice of gender is mainly cosmetic. Only the romances are likely to be influenced, it would be surprising if all the companions and characters encountered were bisexual or playersexual: aka, charming an elf by playing a dwarf and vice versa.

Baldur's Gate 3: History

The histories are used to offer 2 additional Skill masteries to your character. There is also a good chance that they will open up additional dialogue options between now and the game's release. It's a complex topic and there is no perfect choice. You must at least make sure not to choose a background that gives redundant bonuses with your race or class (eg: Sailor Elf).

Persuasion is a fantastic skill offered by quite a few histories. If you're not too confident, choosing Noble or Guild Artisan probably isn't a bad choice. Other options are Criminal, Charlatan and Street Kid if you are playing a Rogue or want to take over his role with another class.

LEATHER & GRATTOUILLE

As mentioned above, we find a bit of the DNA of Larian Studios' previous games in Baldur's Gate 3. For players who like secrets and discovering treasures in the environment, this is a nice little extra. However, this can be frustrating when you have not understood all the subtleties of the thing. We will fix it.

Where to find a shovel?

This mundane object can be collected in many locations in Baldur's Gate 3, you should quickly find one without difficulty if you pay a little attention. The very first shovel in the game is found a little after evacuating the wreck of the Nautiloid, at the start of chapter 1.

It is placed on a mound of earth, at the following coordinates: X:203 Y: 385. Here is an overview of the map of the region with its position. It is possible to miss it by passing to the other side of the rock, or by not paying attention. This also serves as an intuitive mini tutorial, as you are then invited to dig into the mound of earth, in order to reveal your own treasure.

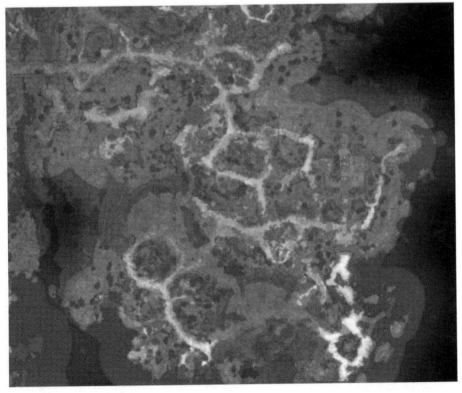

How do you know where to dig?

There are at least 3 methods to know exactly where to dig to reveal treasure. They're not perfect, but by combining them you shouldn't miss out on many treasures.

- The first, which is the simplest, but also the most random, is to approach the area, and make a Perception roll with your characters. A die will be thrown above the heads of your characters and will notify you if you have succeeded or not. If successful, the mound of earth to dig into will appear and you can simply click on it. Unfortunately, it regularly happens that everyone fails, which often gives you the choice between loading the game to start again, with the Assist spell, or returning to camp, to take new companions who have yet to make their Perception roll. .

- This doesn't work for other hidden items, like switches, but for treasures, the shovel is your friend. Add it as a shortcut to the right of the action bar. Use it to click on the ground, all over the area, after a failed Perception roll. By clicking in the correct area, you will still reveal the treasure.

The last method consists of recruiting Scratch, the dog who is in the woods, a little before the ravaged village. By sending him to your camp, then taking care of him a little, you will become his new master. After a long rest or two, a new power will appear in your hotbar: Scratch Pet. This allows you to summon it. Its usefulness in combat is very limited, but don't worry, upon death it is simply unsummoned, and you can recall it after a short rest. His real interest is that he will start barking, and he will guide you to hidden treasures. He's much more helpful than Zelda's dogs. Thanks to it, you will know that there is a hidden treasure in the area, as well as its approximate location.

INGOTS

Baldur's Gate 3 is a fairly complex game, with thousands of items to collect, and it's very easy to miss useful things, or sell them by mistake. For example, the Infernal Iron of Karlach is a good example.

What are ingots used for?

Unfortunately, bars are currently of no use (as of August 30, 2023), although this may change in the future. They nevertheless have a significant value, especially for silver and gold bars. You should still collect them and sell them to merchants, in order to reap a nice jackpot.

A crafting system, like alchemy was planned, but this feature was removed from the game, since it was not ready for release. It may be added later. The ingots lying around everywhere are traces of this aborted system.

Smithing related quests

- Contrary to what one might initially believe, the quest Complete the Master Weapon does not ask you for ingots.

- The Adamantium Forge quest doesn't need ingots either, it requires extracting raw Mithral Ore from the deposits. These minerals are useful, so don't sell them. This specific quest allows you to craft 2 items of your choice.

- We mentioned it above, but Dammon asks you for Infernal Iron, which is not available in ingot form. You only need 2 pieces, the whole part. You can also sell the rest. If you cannot forge equipment, Dammon will offer you new items in Chapter 2, then in Chapter 3. It is better to keep him alive, if possible.

ORDER OF ZONES

Unless you complete Baldur's Gate 3 several times in several different ways, or load your previous saves to do otherwise, it's difficult to realize that certain events are happening automatically in your absence. This means you're missing out on important opportunities, and some of the game's content. This is unacceptable for any perfectionist worth their salt! Spoiler alert.

To this, we must also add difficulty. The game will not hold your hand, and it is possible to kill yourself on a boss that is much too powerful for your current level. It's not too bad normally, but to complete the game in Honor Mode, you have to choose carefully what to do and when, while accumulating experience points and equipment. Consult the interactive map if you are lost, or afraid of missing something.

Prologue

The tutorial aboard the Nautiloid is as short as it is linear. The only thing not to be missed is the release of Shadowheart, who is locked in an Illithid capsule. If you enter the gateway, the boss room with Commander Zhalk, it will be too late to do so.

Chapter 1

The beginning of the first chapter is the phase during which you must pay the most attention to your destination and your actions before moving on, because of your low level and the many events available. It's counterintuitive with the urgency of the situation, but consider using Long Rests at the camp fairly regularly, since many important events will take place there. If you try to do everything without sleeping, you will miss important things.

Start by recruiting the starting companions before setting off on your adventure:

- Recruit Shadowheart on the beach.
- Face the Brains on Legs, then recruit Astarion to the west of the wreck.

- Help Gayle get out of the teleport portal northeast of the nautiloid wreck.

- Free Lae'Zel from his cage a little further north.

Your first real choice of destination will present itself, here is the order we recommend:

- Explore the exterior, then the interior of the Damp Crypt to the east to unlock Wither.

- Join the Druid Grove by participating in the battle at the entrance against the goblins. Recruit Wyll, then complete all possible quests inside except the Theft of the Sacred Idol and the murder of Zevlor or that of Kagha. Freeing Sazza the Goblin is a good opportunity to go out after finishing the rest.

Explore the central area of   the region, and the Ravaged Village but above all, do not cross the wooden bridge to the West, leading to the goblin base. Crossing it will trigger events that will cause you to miss content.

- You can cross the destroyed stone bridge to the north to recruit (or kill) Karlach. But watch out for the Gnolls along the way.

- Recruit Scratch and Shovel. Also get your hands on Thay's Book of Necromancy.

- Complete the burning Waukyne Inn quests in the northwest to begin Saving the Grand Duke. If you go to the goblin base first, the fire will be over, and its occupants will already be dead.

- Find Rugan then decide what to do with the goods before going to the Zentarim hideout.

- Face the Owl Bear at around level 3, and especially, before crossing the bridge. Don't kill the owl bear if you plan to recruit him.

- After saving the occupants of the inn and killing the Owlbear, you can venture into the goblin camp to continue the story. Defeat the partying goblins for a chance to recruit Owlbear later.

- Consider completing the grove quests before deciding whether to help the goblins attack it. This will determine whether you recruit Halsin or Minthara during Chapter 2.

- Explore the swamp to the southeast and see Auntie Ethel in the Overgrown Tunnel around levels 3 to 4.

- The Spider Nest and the Spectral Spider Matriarch boss can be faced from level 3 to 4 as well.

It is not recommended to venture to the northwest corner of the map, where the dragon and Githyanki patrol are located before level 4-5 if you intend to face them.

Do not enter the Mountain Pass area, this will take you to Chapter 2 automatically. This can trigger the closure of the Druid Grove due to the Bramble Ritual, or its destruction by the goblins.

- You can begin exploring the Dark Depths from around level 3 to 4. Progression in the area is relatively free, with a fairly constant level of difficulty. The important thing is not to take the ship to cross the lake. It's best to save the crossing for the end of the chapter, after you've completed everything else. You should be level 5 to 6. When you arrive on the other side, the Free Nere quest begins, it has a limited time.

- After saving Nere (or not) and probably killing him, you can attack the last challenge of chapter 1: the Adamantium Forge north of Malforge.

- Once Grym is killed, you can choose to move on to Chapter 2 by taking the passage back to the surface, or via the Mountain Pass.

Chapter 2

If you've completely cleared Chapter 1, you should be able to tackle the majority of the fights in Chapter 2 straight away. You can even return to areas from the first chapter. You must especially be careful not to trigger the events marking the final confrontation and the end of the chapter, by entering the passage at the bottom of the Gauntlet of Shar. Doing so will prevent you from returning to previous areas.

- Join the Inn of the Last Glow and/or obtain the Moon Lantern from the Drider Drow. Be careful, a big fight that could go wrong awaits you in the inn. If the Priestess of Selune is killed in the inn, many important people will die.

- You can explore almost the entire area, this includes the Cursed Shadowlands, the Highlune City Ruins, the Healing House, the Highlune Towers (except the roof), the prison, the Oubliettes.

- You will have to help Halsin lift the curse via Art and Oliver to recruit him, or you will have to free Minthara from prison.

- The Mountain Pass leads you to an optional area, with Lathander Monastery and the Githyanki Nativity Scene. These areas are very profitable with a legendary weapon, but there are several choices that can lead to the death or desertion of Lae'Zel for example.

- After exploring the rest of the chapter, you can enter the Mausoleum and Gauntlet of Shar. It is advisable to take Shadowheart. By completing the Gauntlet of Shar, and deciding what to do with Nightsong, you will trigger the final battle of the chapter as well as Jaheira's fate.

- Go to the Towers of Hautelune, then on the roof to face Ketheric.

- End the chapter by facing Ketheric a second time in the depths of the Illithid Colony.

- After achieving victory, you can resume your exploration or move on to Chapter 3 by taking the route to Baldur's Gate.

Chapter 3

The beginning of the chapter includes some obligatory events, that's normal.

You are limited to the outskirts of the city at the moment. You can explore and complete local quests. Crossing the bridge leading to Dracosire Rock will trigger a mandatory event, the Coronation of Gortash, before giving you access to the second part of the city.

We strongly advise you to consult our Orin Hostage guide before proceeding further, in order to choose which approach you prefer. Otherwise, you risk being deprived of a companion for the majority of the chapter.

If you have completely cleared the first two chapters, your party should be level 10 and above, which theoretically allows you to explore almost everything. As with Chapter 2, as long as you don't start the final confrontation. There are, however, some major quests requiring you to perform actions in the correct order:

- Find the submarine and free the Iron Throne prisoners before destroying the guardian golem factory and killing Gortash. At least, if you want to save the hostages and the Grand Duke.

- Kill Gortash before going to the Murder Court and Temple of Bhaal if you want to rescue Orin's hostage.

- The House of Sorrow and Cazador's Castle both include a companion quest and at least one difficult fight.

- Raphael's House of Hope can be considered the ultimate challenge of the game It is an optional area, but you cannot leave it without defeating the boss. It is advisable to end there.

- Check out our Legendary Items guide so you don't miss out on the best rewards in this expansive chapter.

You can start the final battle by taking the passage to the north, in the ancient city north of the sewers. You must have eliminated Gortash and Orin. The rest is linear until the end of the game.

BEST SKILLS

Even for a veteran of Dungeons & Dragons 5th edition tabletop, or an expert in the first games of the license, it is not easy to determine which are the best skills to choose. The reason being that it greatly depends on the situations that await us in Baldur's Gate 3, and the direction of the gameplay, as well as the story. A bit like on the table, depending on the context, a skill can be either vital or useless. But after dozens of hours on early access, we have a solid idea of   what this looks like, at least for Act 1. The trend should continue in broad terms. We will also explain to you how it all works to help you understand this list, and choose for yourself if necessary.

Baldur's Gate 3: Diversify your skills & don't neglect your companions

Before we get to the heart of the matter, let's clarify that Baldur's Gate 3 involves a group adventure, not a solo adventure. Your main character is not expected to know everything, or to have to choose one way to play and neglect all other options The companions you will choose to accompany you have their own skills, and they can also lead the dialogues for you if necessary. You can also order a companion to stay at camp, in order to bring another one more suited to the situation at hand.

This way, you can have the majority of skills in the game available. That being said, it's useful to have some skills on as many characters as possible, and others are particularly important to have on your protagonist specifically. We will therefore list them in the order of importance that we give them. This is obviously subjective, but it's based on our gaming experience, as well as how often we use them.

Please note that the skill choices available for each class vary. You may be required to choose a specific Origin that offers the skills you desire.

Baldur's Gate 3: What is skill mastery?

Let's open a second parenthesis to clarify that all characters can attempt all or almost all actions, but that they have a bonus or a penalty depending on the characteristic associated with them. For example, Strength for Athletics, or Charisma for Persuasion. Mastering a skill will grant a bonus (minimum +2) to your die roll, which means your chances of success will be higher. This bonus increases as you gain levels, and it can go up to +4, it can also be multiplied by certain class bonuses such as that of the Dodger or the Bard. By combining all of these elements, this greatly increases your chances of succeeding in your rolls for this skill. What follows is therefore the list of skills that it is best to choose to master, either by selecting them from the list, or by choosing a race or class that masters them automatically.

Baldur's Gate 3: Skill List

Here we list the skills starting with those we think are most useful:

❖ Skills - S tier

The skills in this category are fundamental to fully enjoying your adventure, to do without them is to shoot yourself in the foot, both in terms of difficulty, but also in terms of content explored.

Perception (Wisdom): Without a doubt the absolute queen of skills. You will use it very regularly, whether to detect traps, ambushes, small details, secret passages or hidden treasures, or even elements that can launch new quests. You'll literally miss out on dozens of opportunities, and walk into tons of death traps without it. The positive point is that the game tells you when a perception roll is made, and when it is failed. In general, all members of the group are allowed a roll, which often allows you to succeed at least one (but not always). This also allows you to rely on another character like Astarion. Having to load the game every time a perception roll is failed is unbearable in any case. Elves have this skill by default, and it is useful for all characters. Rogues, Bards, Rangers, Druids and Clerics are especially suitable for mastering this skill.

Persuasion (Charisma) : Another omnipresent skill, this time in dialogue. You can get yourself out of almost any situation with persuasion, and it allows you to easily steer the story in the direction you want by rallying NPCs to your cause. This is by far the best skill for playing someone kind and relatively peaceful, although it can also be used to achieve evil goals. Let us also mention the fact that this skill is by far the most useful on your protagonist (even more than Perception), since you will very often be required to use it on your companions so that they come around to your opinion, or that they reveal their secrets to you, which by extension helps your relationship progress. The only reason this skill isn't #1 is because you often have alternatives, whether it's intimidation or using the larva, or new chances to persuade someone later. Although fundamental for the protagonist, it is Bards, Sorcerers, Warlocks, and Paladins who will master it best with their high charisma score.

❖ Skills - A tier

The skills of this tier are very useful, and it is definitely better to have them, but it is not comparable to the previous ones.

Deception and Intimidation (Charisma) : Alternatives that are a little rarer than Persuasion, and can have negative repercussions on occasion, which explains why they are both put together and lower in the ranking. Blatantly lying to a guard and intimidating a goblin is no problem, but on your companions it can cause problems. Since it's Charisma-based, the same classes have the advantage, but it's still a good investment for brutes like Warriors and Barbarians, as well as tricksters like the Rogue.

Sleight of hand (Dexterity) : A formidable skill whose use is quite rare during dialogues, but which will be used extensively by your Dodger, your Bard, or the character who replaces them in the group. It is used to rob NPCs and merchants of all their gold and goods without them realizing it, which allows you to become immensely rich and well equipped very quickly, especially if you abuse the loadouts in case of 'failure. Its use does not stop there, Sleight of Hand is also used to pick locks and defuse traps. Nothing is more frustrating than not having access to the contents of a vault, right? Any class with a good Dexterity score is suitable, and there are many of them. The reason this skill is not in S Tier is that you have quite a few alternatives. There are often keys for doors and trunks. Doors can also be destroyed without repercussions (but not chests). Traps can be defused from afar with a projectile, a spell, or any object. Larger chests and doors often have special mechanisms that do not use Sleight of Hand.

❖ Skills - B Tier

Useful skills that will regularly help you in one area or another. It's good to have them on hand since they will make your life easier in general and provide you with opportunities on occasion:

Investigation (Intelligence) : This skill is used to objectively analyze the situation and to unlock additional, easier or more profitable solutions. For example by giving you the solution, or by allowing you to use another skill, with an advantage on the die roll. Only the Magician will be able to fully exploit it by default, but other classes such as the Thief and the Bard can take care of it.

Arcana (Intelligence) : You will regularly come across magic, strange devices, spells, illusions or curiosities whose use or nature can be explained with the Arcana which will often open up new solutions or make it possible to use them. Once again this is normally for the Magician, but failing that, the Warlock or Sorcerer can take care of it.

Insight (Wisdom) : Another skill that shines during dialogue. It allows you to objectively determine if someone is lying to you or trying to trick you, which helps you make the right decision. This sometimes opens up additional dialogue options, especially with your companions, but this is rather rare. Once again, it is more for classes with wisdom by default, like Cleric, Druid and Ranger, even if it is a good complement for a character who is a master of dialogue.

Stealth/Stealth (Dexterity) : For the moment, Baldur's Gate 3 uses a lighting management system as well as fields of vision displayed in color to manage stealth. Discretion therefore only comes into play when the character is likely to be spotted passing through a field of vision. By playing well, or by using spells like invisibility, you can largely do without it. This is used in practice to assassinate groups of enemies with his Rogue and his sneak attack without being spotted, it is then formidable and justifies the fact that this skill is not relegated to the bottom of the ranking, because the other classes are not. don't really use it. This will perhaps change with the evolution of early access.

Athletics (Strength): Used to perform physical feats such as pushing an obstacle blocking the path, or freeing someone trapped under rubble. However, this is very rare. In practice, it is mainly used to determine if your character is going to crash like an idiot after a big jump below. In case of failure, damage is suffered and he is immobilized for a few seconds. As it is risky, it is best to avoid such maneuvers in combat.

❖ Skills - C Tier

The skills on this list are either used too rarely or have too little impact.

Training (Wisdom) : The rare version of Persuasion, but intended for animals. As with the skills above, this opens up very interesting opportunities, but its use is also very rare, which explains why it is low in this ranking. Rangers, Druids and Clerics will make the best use of it.

Medicine (Wisdom) : In a world where healing magic and healing potions are commonplace, the use of medicine is limited. We haven't found many opportunities to exploit it. It was very useful in these cases though.

History (Intelligence) / Religion (Intelligence) / Nature (Intelligence) / Survival (Wisdom) : These skills serve more to offer information to the player than anything else. They also sometimes unlock special responses during dialogue, but they have never advanced the situation in a significant way.

Acrobatics (Dexterity) : The type of throw is not reported in-game, but it appears that acrobatics helps avoid attacks of opportunity as well as certain ground effects like grease and vines. Currently, the disengagement leap makes it (too) easy to avoid attacks of opportunity in exchange for a bonus action. For other cases, it is indeed useful, even if there are ways to avoid this manually. We should also expect enemies to abuse area effects on the ground a little less in future patches.

❖ Skills - D Tier

The skills on this last list are useless or almost useless.

Representation (Charisma) : Unless you play Bard, a class currently absent from the game, this skill is almost useless. The instruments found are broken, and it is only a poor alternative to Persuasion, Intimidation and Deception.

BEST EARLY GAME ITEMS

There are a multitude of magical items scattered all over Baldur's Gate 3, some are simply in chests, or with merchants, but others are hidden in a secret location, or linked to certain choices on your part, which makes them very easy to miss permanently. It would be a shame to miss out, since some of them are capable of influencing the way you create and develop your characters. We will list here general objects, useful for almost all classes and all characters, rather than those overpowering for a specific class.

At the moment we are only listing items listed in the first act of the game, which you can obtain relatively quickly, which is the most important thing. The others will be added later. Note that you can easily buy magical +1 weapons and armor from the merchants in the Druid Grove, at the start of the first chapter.

Longflame Sword

- Bonus: 2D6+1 slashing damage +1d4 fire damage
- Source: On the corpse of Commander Zhalk, aboard the Nautiloid, during the tutorial.

Towards the end of the tutorial, when you arrive on the ship's bridge, and the devils confront the mind flayers. On the surface, this is not a fight you can win, but in reality, you can take advantage of the situation to grab the best weapon at the start of the game, a very powerful flaming greatsword. It's best to free Shadowheart then use the Shield of Faith on the Mind Flayer, to help him survive long enough. By finishing off Zalk, you will receive 75 XP, then you will be able to flee before the end of the 15 turns. But be careful, when Commander Zhalk dies, the Mind Flayer becomes hostile, you must distract him with reinforcements, or be ready to activate the console.

Even if you don't plan to use this sword, it will suit Lae'Zel perfectly during the first dozen hours of play.

Amulet of Lost Voices

- Bonus: Allows you to cast the level 3 spell Communication with the Dead.
- Source: In the damp crypt, near the departure beach, after the ship crash. She is found in the heavy camouflaged chest at the bottom of the crypt, after confronting a group of undead scribes.

This amulet is of no use in combat, but it opens many doors for you, since it allows you to extract information from many corpses that you will come across. This often doesn't lead to anything, but sometimes it gives you rewards, quests, experience, etc. This is a tool that you should not forget to use whenever possible.

Headband of Distorted Intellect

- Bonus: The wearer's Intelligence increases to 17
- Source: Found on Ogre Mage Lump after killing him, in the Dilapidated Village.

Certainly the most popular helmet at the start of the game. For Magicians, it's an ability bonus at low levels and for everyone else, it's a massive bonus on quite a few skill rolls (Arcana, History, Religion , Investigation, Nature). It also allows you to have access to additional options during dialogues, requiring you to have a certain intelligence score. As it is very easy and quick to obtain, it is even possible to completely neglect the Intelligence of your character when creating it, then to compensate with this headband.

An easy way to get this helmet, even at very low levels, is to persuade Lump to fight for you. You have to promise him 1000 gold coins and make a subterfuge roll, but he will give you a foghorn in exchange. It can be used to summon him and his teammates during a fight. With their help, you can even kill the fearsome Phase Spider Matriarch quite easily. You just have to let him perish, or finish him off afterwards, deaths don't need to be paid, and you can pick up his crown.

Talisman of the absolute

- Bonus: When the wearer has less than 25% of his testing health points, and inflicts damage, he regains 1d8 life points.

- Source: Found in the Shattered Sanctuary, the goblin base, on the corpse of Gut, the High Priest of the Absolute, after killing him.

This talisman can help a character survive a difficult situation. This is particularly effective on martial classes that have several attacks like Lae'Zel, and even more so with two weapons. They can then recover health quickly, which allows you to heal someone else, or to be more aggressive.

Amulet of Misty Steps

- Bonus: Allows you to cast the level 2 spell Misty Step (teleport 18m) counting as a bonus action (and a spell).

- Source: Sold by Omeluum the merchant in the Myconid colony in Outland.

This talisman will greatly help one of your most fragile characters to survive by escaping from a distance while saving their spells, or one of your fighters to reach the melee instantly by ignoring obstacles. It also helps to reach positions that are otherwise difficult to reach. Suffice to say that you will always find a use for it.

Mysteries revealed

- Bonus: Allows you to cast the level 2 spell Detect Thoughts.

- Source: Found next to the ascender in the Arcane Tower, Outland. You will need a Sussur Flower to unlock it.

Like the amulet of communication with the dead, jewelry besides quite a few options during dialogues. Its effect can even be launched while talking to someone, without having to do so first. This allows you to discern your intentions, discover information, and even unlock new options, by reading your password in your mind for example.

Ring of psionic protection

- Bonus: Forms a barrier around the Illithid larva. The ring's bearer can no longer harness its power, but it also does not gain any power.

- Source: By completing the "Remove the Parasite" quest section from Omeluum the Merchant in the Myconid Colony in Outland. Which then requires you to complete quite a few steps with your friend Illithid and the Arcane Tower.

We do not yet know all the branches of the story of Baldur's Gate 3, nor its ending, but it is a safe bet that to obtain a positive, or "good" result, you must avoid using the larva and let it grow. At least that's our theory. If this is the direction you want to take, this ring is the only real advancement available during the first act of the game.

Callarduran Smoothhands Fetish

- Bonus: Allows you to cast the level 2 spell Invisibility
- Source: In Grymforge, on a corpse located at coordinates X:-610 Y:409. Two Duergar throw gnome corpses. They should be encouraged to loot the bodies before throwing them into the water. You are then entitled to a detection roll to find the ring.

Invisibility is a fantastic tool, especially in Baldur's Gate 3 where stealth is no longer an absolute cheat code. This can make your Thief really hard to find. In addition to saving you spells, this also allows a character as stealthy as an armored Paladin to escape detection.

Grymforge is an area full of powerful items, from armor pieces to boots and jewelry The area is nevertheless quite high level, with a big level 8 boss, which explains why we are not listing the rest for the moment. But don't hesitate to take a look as soon as you have the required power, you won't regret it.

PERMANENT STAT BONUSES

In Baldur's Gate 3, the vast majority of bonuses come from level gains, equipment, and temporary buffs, such as spells and potions. But it's exceedingly difficult to achieve permanent power gains outside of that. And this is especially true for features. Especially if you want to exceed the 20 point limit.

Bonus Character Points

Three opportunities to obtain characteristic points have been listed at the moment. They can allow you to exceed the normal limits of 20 in a characteristic, but they also have some flaws. It seems that using a Respec with Wither causes them to disappear. It is therefore better to be sure before using it.

Chapter 1: Auntie Ethel's Hair

During the confrontation against Auntie Ethel, at the bottom of her cave, during chapter 1, she may ask you to spare her life. For this, it is necessary to have pushed her to reveal her true form in the tea room, by seeking to tell the truth about the deaths of the brothers. Then, during the fight, Ethel's life must be very low, without killing her. She will then offer you an arrangement, letting her escape with the girl, in exchange for her hair, an object conferring a bonus of +1 Point in any characteristic of your choice when consumed. By making a good skill roll, you can also force her to let you have the extra girl. This is a good decision, since Auntie

Ethel comes back to life in Chapter 3, so killing her at this point accomplishes nothing. Be careful though, this will break your oath as a Paladin of Vengeance. Be careful, it seems that Ethel's hair disappears after a long rest or a change of chapter it is not possible to keep it for too long.

Chapter 2: Potion of Eternal Vigor

You must infiltrate the Towers of High Moon, southwest of the cursed lands of chapter 2. On the ground floor is a drow alchemist named Araj Oblodra, who is strangely fascinated by blood. She will also ask you for some of yours for her experiences, which you should accept.

She will also mention Astarion the vampire, and ask you to bring him in to offer him a special opportunity, if he is not already in your party. You will then have to convince Astarion to bite her and drink her blood, which tastes bad, but in exchange you will get a potion offering a massive +2 Strength bonus. It is better to consume it on a character who already has 20 Strength.

Chapter 3: Mirror of Misguidement

Now, go to the slums of Baldur's Gate, to the building in the northwest corner of the map, the House of Sorrow. This shady therapy center actually hides a temple of Shar. It is advisable to go there with Shadowheart to conclude his personal story. We won't spoil the story for you, but after a big fight, you have to open the second part of the temple with the black diamond placed on the altar. You will then find an immense mirror of confusion capable of devouring memories.

Make a save before approaching him, since several skill checks must be made to fully exploit his potential. First, an Arcana roll, to identify its function. Then choose the option to pray, which will trigger a religion roll. Finally, approach to voluntarily offer a souvenir.

You have several options to choose from, including offering the forbidden knowledge from Thay's book, which we advise against, but it is better to choose a characteristic that you do not use, such as Intelligence in the majority of cases, since this will inflict a -2 penalty on you. It seems that deceiving the mirror by inventing memories achieves nothing. Then, if you have followed the previous steps correctly, you will be offered to gain 2 characteristic points of your choice. This is the opportunity to reach a score of 22, or even 24 for Strength (or even potentially 25, even if it doesn't bring anything more).

The good news in all of this is that the ability penalty is a curse. You can remove it with the remove curse spell, possessed by the majority of spellcasters, such as Wizards and Clerics. So it's all good.

POSSIBLE ENDS

Trying to list all the possible endings of Baldur's Gate 3, and what could have influenced them, is far too long and complex, especially without having been able to test everything. Nevertheless, we were able to experience the main endings of the game. There are others that require very specific choices throughout the campaign, which will be added in the future.

This should be obvious, but spoiler alert!

The Emperor, Prince Orphéys and the transformation into Illithid

If you progressed through Lae'Zel's quests, and obtained the Orphic Hammer somehow, you must have been looking into how to enter the prism to free it. This option is only offered to you at the very end of the game, after the failed confrontation with the Infernal Brain. A series of choices is then offered to you:

- Letting the Emperor absorb Orpheys: This will greatly upset Lae'Zel and you will need to persuade her or kill her if she is in the group, as will her Githyanki allies. You can order him to destroy the Infernal Brain at the end, or kill him in order to become the Absolute yourself.

- Free Prince Orpheys: The Emperor will then betray you and join the camp of the Absolute, but in exchange, the Githyanki will join your cause. You will have to kill the Emperor during the final battle. The prince will collaborate with you, to the point of being willing to sacrifice himself and become an Illithid himself. This gives you quite a few possible endings, such as destroying the Absolute, then executing it or not, or betraying it to become the Absolute yourself.

- Becoming an Illithid: It is possible to sacrifice yourself by teaming up with the Emperor, which involves devouring the prince. Once again, Lae'Zel and his Githyanki allies are not going to be happy. The alternative is to free the prince, and then sacrifice himself to become an Illithid. This is by far the most Githyanki-friendly option. New options are also available at the end of the game, such as trying to dominate the Absolute or destroy it, which can push your ally to turn on you (the prince or the Emperor), if not consistent with their beliefs. After the brain is destroyed, you can choose what to do, such as going into exile or staying in town.

- Push Karlach to become an Illithid: To have access to this option, Karlach must be in the group. As she knows she is condemned by her heart, she will offer to sacrifice herself. The rest unfolds in a way quite similar to the previous choices, you can choose to free the prince, or to devour him. You will be able to choose the fate of Karlach at the end of the story.

Best ending: How to get the best possible ending?

Without going into too much detail, what we can consider to be the best ending in general, at least from our point of view. It is possible to prefer certain variations regarding the fate of certain characters. This requires making the following general choices:

- Save the Tieflings in Chapter 1.

- Prevent Isobelle's kidnapping in Chapter 2, then free Chantenuit.

- Kill the chosen ones of the dead gods, then free Prince Orpheys with the Orphic Hammer. Then push him to sacrifice himself by transforming into an Illithid. You must not have become half-Illithid either. After the final battle, let him destroy the Absolute.

- All possible companions were recruited and kept alive, with a positive outcome to their personal quest (Shadowheart, Astarion, Gayle, Karlach, Wyll, Halsin, Jaheira, Minsc). The most difficult thing is surely keeping Karlach alive, we will detail this further below. Some feel that letting Karlach become a Mind Flayer while saving the prince is a better option. This is obviously a question of point of view, and the result is mixed at best.

- You must not have exchanged the Orphic Hammer in exchange for the Crown of Karsus to Raphael.

Bad ending / Evil Ending: Becoming the Absolute

There are several possible variations for this ending, depending on whether you have transformed (or not) into a Mind Flayer, and the allies at your side. But in summary, you make the choice to betray everyone at the last moment, in order to take control of the Infernal Brain, and become the new Absolute. In this case, the fate of each companion no longer has any importance.

Special ending: Dark impulses

You will learn more about the origins of this character, and that of his impulses during the story. This allows for additional scenes, especially when you face Orin. It is then possible to accept Bhaal's gift and become the Ravager, his new chosen one. The alternative is to reject him and die, before being reborn free thanks to the intervention of Wither.

By choosing to receive Bhaal's gift, this results in slightly different lines of text at the end. You can choose to resist your impulses and destroy your brain, or to betray your ally in order to dominate them. Once on the throne, celebrating your victory, you then pronounce Bhaal's name, rather than your own. If you decide to destroy the brain, you know that punishment awaits you, but you can deal with it in different ways: by committing suicide, by being imprisoned, or by planning to face the god himself.

Raphael

If you have decided to exchange the Orphic Hammer in exchange for the Crown of Karsus, you are entitled to an additional scene at the end of the game, during which Raphael thanks you. He is about to unify the 9 hells and become their absolute lord. He then intends to continue his conquests, and one day attack the primary plane. You have resolved the current crisis, but an even bigger crisis awaits you in the future.

Game Over

There are many parts of the game where your bad decisions lead to an instant game over, prompting you to load the game. For example, when you don't deactivate the Blood of Lathander solar spear in time, or when you challenge a certain lich to prove to you that it has the power to kill someone from a distance. Let's also mention sleeping with an Incubus in the House of Hope, if you fail certain saving throws, he devours your soul. There is, however, a key moment in the game that is considered a legitimate ending, as the end credits then appear on screen.

If Gayle is in your party, when you reach Ketheric Thorm's Refuge, deep within the Illithid Nest, you can encourage him to detonate the Orb of Destruction in his chest The titanic explosion then instantly destroys the three chosen ones of the dead gods, as well as the infernal brain. Of course, your party dies in their company, but it's fun to see their surprise. Unfortunately, it's a bad ending, since all the people contaminated by the larvae turn into Illithids who will sweep across the Sword Coast. We can then see Elminster lamenting that he pushed you to make this stupid choice (and literally did nothing about the story too).

Fate of companions

One of our regrets about Baldur's Gate 3 is that it doesn't have a cutscene for each companion, or an image with text, as you usually see in CRPGs. This was a strong element of Baldur's Gate 2, and it was interesting to see the different possible fates of each companion depending on your actions. We only list here the companions who are entitled to a special passage at the end of the game, at least, if they have survived until then.

Karlach

Without the slightest doubt the most tragic end, since she is condemned to transform into a Mind Flayer, or to see her infernal heart overheat. Several elements seem to indicate that content was cut regarding Karlach, and a solution should have been available to repair or cool his core. Perhaps a future DLC, or an Enhanced Edition will remedy this. There are three possibilities for our favorite barbarian:

- The first is simply to die in your company on the primary plane.

- The second is to return to hell, to Avernus, in the company of Wyll, if he has become the Blade of Avernus, rather than the new Grand Duke. This also involves breaking his pact with Mizora, and saving his father in the underwater prison.

- The last option is to accompany Karlach to Avernus yourself, if you are a couple.

Astarion

If you have not completed the vampire and Cazador's personal quests, or he has rejected the ritual, he begins to burn in the sun during the ending scene. He is then forced to take refuge in the darkness. Otherwise, if you have successfully completed the ritual, he can live even in broad daylight, but his nature has also become much more evil.

Lae'Zel

It is by far the companion which has the greatest number of possible variations. Her destiny changes greatly if she decides to remain faithful to her queen, or if she rebels, for example. The second important factor is the fate of Prince Orpheys: He may survive and help you in battle, but your protagonist has then become a mind flayer. In this case, Lae'Zel will assist him in freeing their people. An alternative is to trick the prince into becoming a mind flayer, and he asks you to execute him after the battle, which you may or may not do. Either way, you can let Lae'Zel go, in order to fulfill her destiny, or you can persuade her to stay with you. It's much easier if you haven't become an octopus, and you're having a romance with her.

Gayle

If he's in your group, the idiot Magician of the group can let out a little "I'M ATOMIC!" to the Infernal Brain, only to cut short the final battle with a spectacular explosion, which leads to his own rather disastrous end too, as well as his death, incidentally.

Otherwise, Gayle will try to fish out and repair Karsus' crown to offer it to Mystra so that she can heal him (or not). At least, if you didn't promise it to Raphael in exchange for the Orphic Hammer.

For these different options, you must have fed Gayle's orb with magical objects, have met Elminster at the start of chapter 2, then have gone to read the annals of Karsus under the Magic & Witchcraft store. Elminster will then give him a new appointment with the goddess Mystra, at the temple of the lower districts. We must then pray at the statue of Mystra.

Shadowheart

It seems that the final dialogue with Shadowheart suffers from a bug, and it does not always appear after the end of the game, which was the case for us. His ending depends greatly on whether he is redeemed or not. For the best possible ending, she must have spared Chantenuit, then spared her parents in the Temple of Shar. She will then announce her intention to build a house for her parents. If you have a romantic relationship, then you are welcome to live with her.

Wither

It's never said openly, but as suspected from the start, Wither appears to be the ancient god Jargal. He gets a post-credits scene, if you have a good ending. He mocks the three dead gods and their ridiculous plan to steal the souls of the victims transformed into Mind Flayers.

Elected of the absolute

There is no end to keeping the chosen ones of the Absolute alive, or allying with them as far as we know. Circumstances and their beliefs always force you to kill them, or die at their hands.

ASTRAL LARVA: SHOULD WE EAT IT?

The very heart of Baldur's Gate 3's story is the Illithid larva inserted into your brain in the game's opening cutscene. Discovering that it grants you powers similar to those of the infamous and fearsome mind flayers, you won't You might be tempted to use them, but at what cost?

We will review the 3 elements related to the thing in this guide: the use of powers, the consumption of larvae in a jar, and for players more advanced in the story: the astral larva.

Use the powers of the Illithid Larva

You will quickly have the possibility of using and abusing the powers of your larva, in order to influence and manipulate your enemies, as well as your companions.

It often comes in this form: [Illithid][Wisdom].

And our answer is that you should use them if it fits your character's roleplay. There are absolutely no negative long-term repercussions. In the worst case, you risk angering one of your companions. In this case, load the game. There is no fear of being stuck, and not getting the good ending, because of the use of your powers.

Consume Mind Flayer Parasite Specimens

After a few hours of play, you will begin to regularly find larvae in jars, whether in the environment, or on the corpses of the Absolute's followers. Goblin leaders are good examples of this. You cannot use them immediately, you must wait until you have a dream of the Guardian Angel, during the night, who invites you to do so. Subsequently, you can use them on your protagonist, as well as on some of your companions. You will need to persuade others to use them.

After consuming one, it will bring up a new interface with your brain, and new talents to unlock. They're often mediocre, but some are really great, like Flying, Skill Mastery, or Extra Action.

Unfortunately, this is also of no consequence, even if you unlock a lot of bonuses. Some of your companions may criticize and disapprove of your actions, but that's it It is kind of a shame. This is one of the missed opportunities of Baldur's Gate 3.

Use the Astral Larva

By starting chapter 3, your Guardian Angel will finally reveal his true nature: a powerful mind flayer nicknamed the Emperor. After the battle against the Githyanki guardians, he invites you to further your transformation using a very special larva that will make you a true hybrid. Your appearance won't change too much, aside from a few purple lines on your face, but it will unlock new powers.

The 5 powers at the center of the brain are then automatically unlocked. If you manage to convince your companions to use the astral larva, you will save 5 parasites in a jar per person.

It's sad to say, but apart from a few lines of dialogue, and a little disapproval from certain companions, this has no consequences. Becoming a half-Illithid doesn't end the romances, and it never stops you from choosing the ending you prefer.

As you will have understood, this is one of the flaws of the game. The developers did not dare (or did not have the time) to integrate real consequences linked to the use of Illithid larvae and powers. So use it as much as you want. Please note, however, that it is possible (even easy) to finish the game in hard mode, without ever using it.

NON-LETHAL WEAPON

While old RPGs tend to make the hostility of characters and guards permanent, this is not the case in Baldur's Gate 3. You will often encounter temporary states of hostility, which allows you to flee or knock out enemies and come back later without the situation immediately degenerating. Using non-lethal attacks is also a great way to unlock additional opportunities and dialogues.

How to carry out non-lethal attacks?

It's very simple, but there are important restrictions. Select one of your characters, then click on the "Passives" tab on its action bar. This will display a gray icon, with a hammer symbol hitting a humanoid head. By clicking on it, your character will automatically knock out enemies instead of killing them, when they drop to 0 health points. Instead, they will keep one, and be knocked out permanently, or temporarily, depending on the case. But this only works with melee attacks (fists or weapons).

- Using ranged attacks, spells or objects (crates, bombs, vials, environment) remains deadly. It must be said that inflicting non-lethal damage with a Fireball is not easy.

- This does not work on undead and constructs. It also seems that in some cases it automatically kills characters with whom only conflict is possible, and for whom there are no repercussions for death, such as shapeshifters.

Advantages of non-lethal attacks

In addition to avoiding killing someone, which may be necessary to complete a quest, or avoiding losing a merchant, there are various nice bonuses.

- To start, you can search stunned people, and steal all their equipment, as if they were dead.

- Knocking someone out still gives you the experience their death would have given you.

- If you are playing a Paladin, this allows you to avoid breaking your Oath, when facing innocents or guards for one reason or another. This is undoubtedly the most role play reason. This prevents you from becoming a Forsworn following a failed persuasion roll.

- This is quite a feat, but if you save the game after knocking someone out, since you're loading said save, you can finish them off afterwards. This will earn you experience again. It's a bit of a pain, and mostly inglorious, way to get tons of extra experience early in the game.

In summary, there is no reason not to activate non-lethal attacks on all your characters, it only brings benefits.

Who to keep alive with non-lethal attacks?

We will not pretend to know all the situations and opportunities during which non-lethal attacks will give rise to a special situation, but we have listed at least two. We do not count the experience farming exploit. Spoiler alert.

- When you encounter Minsc in the sewers, after the bank robbery, you must use non-lethal damage to knock him out. This allows you to finish off the agents of the absolute, then try to rally him to your cause in the company of Jaheira. This is by far the most notable example.

- In Raphael's House of Hope, if you evacuate with Hope, his sister will appear during the final fight. But she rallied around Raphaël. Sparing your life changes the dialogue with Espérance a little, who will thank you for it.

Silver Sword of the Astral Plane

Larian Studios has put effort into making Baldur's Gate 3 consistent, and that sometimes involves sacrifices. Players figured out how to get what could be considered the best legendary weapon in the game, as early as Chapter 1, instead of at the end of Chapter 3, using the right approach. For comparison, the most accessible legendary weapon, the Blood of Lathander, is not normally found until halfway through Chapter 2. Thanks to Asheryth for the images.

How to get the legendary weapon in Chapter 1?

We might as well warn you, the method below abuses the game's mechanics. It is not impossible that this will be corrected sooner or later in a patch.

To start, you need a Warrior of at least level 3, with the Warmaster Subclass, and the special ability "Disarmament". This can't be Lae'Zel, unless you chose her as the original character. The reason being that a script will make him temporarily leave your group when approaching the area concerned. You can respec with Blight specifically for the occasion.

Sneak up to the Githyanki patrol in the northwest corner of the Chapter 1 map. They are accompanied by a large red dragon, and they trigger a cutscene, they are hard to miss. You must detach your Warrior from the rest of the group.

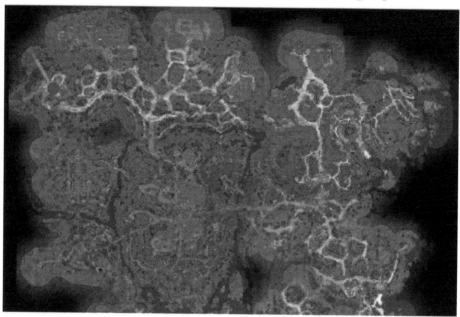

Go through the back of the area, using an Invisibility Potion, spell, or similar effect on your Warrior. Also remember to make a backup, the chances of success of the operation are quite low. They often hover around 10%, which will require you to try several times unless you are lucky.

Use the disarm ability on Kith'rak Voss, the Dragon Rider. You can do this in melee or with a bow, depending on your stats and preferences. Pickpocketing doesn't work, since he has the legendary weapon equipped.

If successful, the Silver Sword of the Astral Plane will fall to the ground. But in all cases, the dialogue will begin automatically and it will generally lead to a fight with the patrol. This is a much more profitable option for you, in terms of experience and loot, even if it is not the priority in this case.

After the dialogue, Kith'rak Voss automatically leaves the area with his dragon, forgetting his sword on the ground (if successful), what a fool! The fight then begins. Pick up the sword with your Warrior as soon as possible. He will then be killed by the patrol, unless you intervene quickly with the rest of the group. Letting him die is not a problem.

Return to camp, speak to Withered again, then pay for the resurrection of your deceased Warrior when needed. He will then reappear at the camp with the precious sword in his possession.

COMPLETE QUEST WALKTHROUGH

CHAPTER 1

Commandant Zhalk

The beginning of the adventure aboard the mind flayer ship can take a different turn if, against all odds, you win the huge battle on the bridge, in order to obtain a two-handed sword, the burning blade and a lot of 'XP. Find out how in this guide.

The introduction/tutorial aboard the Nautiloid normally asks you to reach for the ship's console, which triggers a cutscene and the transition onto a beach. It must be said that the combat taking place on board is much too high level for a bunch of level 1 beginners, who are struggling to survive, and logically, you should have no chance of winning it. At least that's the theory, and players have managed to achieve this feat with a lot of preparation and a very specific strategy.

❖ Two-Handed Sword: Longflame Blade

If everything goes perfectly, your characters will obtain level 2, several hundred gold coins and a Longflame two-handed sword 2d6 +1 + 1d4 fire damage, perfect for Lae'Zel or a martial protagonist.

❖ Strategy

- The first important step is to free Shadowheart from her capsule, so that she can join your group. It is absolutely vital to give you a chance of victory. To do this, search the next room to obtain the console activation rune next to it.

- You can get quite a bit of experience before the fight by going to the capsule that gave you the runestones. You can attack the protruding foot to make shadows appear that you can kill. However, this requires a lot of patience and practice.

- When you enter the Nautoloid cabin and the fight begins between the two factions (Mind Flayer vs. Cambions), victory depends on maintaining a precarious balance between the two camps. We must manage to keep the Mind Flayer alive, so that he kills the leader of the Cambions, Commander Zhalk, then the two powerful Cambions who arrive as reinforcements. If the Mind Flayer dies, the Cambions will slaughter you. And if the Mind Flayer ends up attacking you with too much life remaining, it will also kill you in one turn.

- A good dose of luck is required to win this fight, we advise you to save the game before opening the door and starting the fight. Also, don't hesitate to make quick saves during combat, if events take a turn in your favor. You can charge at the situation getting out of hand, hoping to have better luck on the next attempt.

- Start by casting Shield of Faith on the Mind Flayer with Shadowheart, to improve its survivability. He will really need it.

- Kill all the small enemies in the room as a priority.

- Then concentrate your attacks on Commander Zhalk, while advancing to the back of the room, in order to protect yourself from reinforcements. The next step is to find the right time to spawn the additional Cambions, so that they occupy the Mind Flayer. If he finds himself without enemies, he will then become hostile, which has a high chance of killing you. Cambions appear when you approach the console, at the back of the room, or after a few days. It is important that they appear and reach the melee before Zhalk dies.

- Shadowheart's Tracer Bolt spell is fantastic for damaging and killing Cambions and Mind Flayer, as it will provide you with advantage on your attacks afterwards.

- Your objective should ideally be to kill the 3 Cambions, and to have the Mind Flayer below 10 health points, in order to have a chance of finishing him off quickly (remember to save towards the end).

- If you really can't kill the 2 additional Cambions and the Mind Flayer, it may be wise to collect the loot from Zhalk, who holds the Longflame Sword, and flee.

❖ Cheese

There are different methods, not always very glorious, to improve your chances of winning the fight:

- Keep a character in the previous rooms, next to the healing console ("Us", usually). His task will be to use it each turn to heal all your characters, which will also restore part of their special skills, as if they had rested. It will even bring them back from the dead. It is difficult to lose in these conditions. This allows Lae'Zel and Shadowheart to tank the two Cambions who arrived as reinforcements, while you let the Mind Flayer deal with the commander.

- You can also camouflage yourself, and use the mage hand to throw explosive barrels or their equivalent at the Cambions. By hiding in the upper level of the room, and by abusing stealth, you can also finish off the Cambions with a bow or crossbow, even if it is excessively long, and you only have 15 turns to activate the console, under penalty of game over.

Saving Sazza the Goblin

When visiting the Druid Grove, you will come across a goblin in prison, who needs your help. Helping him escape with this quest can be a good idea, with great rewards as we will see in this guide.

Baldur's Gate 3 gives you quite a few opportunities to switch sides, or infiltrate, and the goblins are no exception. Getting a goblin out of prison can be a good plan to get into their base without having to face everyone.

❖ Saving Sazza the Goblin

After the battle in front of the Emerald Enclave, you can enter and reach the prison Inside is Arka the Tiefline, who threatens to execute Sazza in her cell with a crossbow. You can let her get shot down, but it goes without saying that this will end the quest prematurely. Oath of Devotion Paladins also risk breaking their oath.

- You can convince Arka to leave with a Persuasion or Intimidation roll.

❖ Free Sazza and escort her

You can then talk to Sazza, and agree to help him escape. But before that, we advise you to go and clear the ground, since you will have to escort him. Taking her through the populated areas of the grove would be suicide, and even in the best case scenario you would have to kill everyone, which is not a good idea in this context.

Before freeing Sazza, go down the path along the cliff, not far from his cage. You will have to succeed on a Perception roll to reveal a hidden door.

Passing through this door, you will enter a tunnel with traps: animal statues, which shoot lasers. You must deactivate them by interacting with the blue runes on the walls. Using stealth allows you to avoid being targeted by statues. You will have to deactivate them to prevent Sazza from being transformed into a hot dog.

You will also come across a group of goblins who will attack you on sight: kill them. If you have Sazza in the group, they will still attack, which will put her in danger.

Once all these dangers have been cleared, you can go and free Sazza, and escort her through this tunnel. You must reach the opposite door, which will take you directly outside the druid grove. The other door leads inside the grove, make no mistake.

Once outside, Sazza will meet you in the goblin camp, but her quest is not over.

❖ Accompany Sazza to the goblin camps

To find Sazza, cross the ruined Villa (watch out for the goblin ambush), then cross the wooden bridge going west. Sazza awaits you at the entrance to the small goblin camp blocking the way. Talk to him to progress:

- The advantage of having Sazza is that she will automatically convince the sentry to let you pass. Otherwise, you must succeed at skill rolls, use your Illithid powers, or massacre its occupants.

- As you continue on your way, Sazza will guide you to the entrance to the goblin party, then into the base itself. Remember to talk to her every time you see her to move her to the next step.

- Finally, you can find Sazza at the back right of the base, talking to Minthara.

❖ Sazza, Minthara et récompense finale

Sazza will report to Minthara the drow, and reveal the location of the Druid Grove. You can intervene in different ways, depending on how you plan to direct the story:

- You can expose Sazza's lies, revealing that you had to save her and do all the work, and that she is now trying to take credit for you. She is then executed by Minthara.

- You can also leave her quarter of an hour of glory to Sazza, who is then spared.

In any case, you will receive the quest reward: an assassin's dagger.

Things then get complicated, since Minthara wants to organize an attack on the Druid Grove, ordering you to participate. Your options are then limited:

- Refuse and confront Minthara and her goblins immediately.

- Pretend to agree, and betray the goblins during the attack.

- Accept his offer, and help him destroy the druid grove, which will have serious repercussions for the future. But this will allow you to start a relationship with Minthara and recruit her.

Find your possessions

One of the many charms of Baldur's Gate 3 is that sometimes failing a skill roll is more interesting than succeeding. The quest Find Your Possessions is a good example of this.

❖ *Start quest: Find your possessions.*

Enter the Druid Grove, and advance through the area. You will eventually come across Mattis, a Tieffelin child in front of a stand at the top of the stairs. She praises the merits of a lucky coin, asking you to choose heads or tails. You will always win. You can also quickly report their scam attempts and take the conversation in different directions. It is advisable to show kindness for the future.

A perception roll will take place at the end of the cutscene:

- If you fail, you will realize that someone has stolen from you, which will launch the quest Find Your Possessions.

- If successful, you catch the kid who was trying to pick your pockets, and you are free to treat her as you wish. Once again, it is better to show kindness if you want to be able to unlock other opportunities later. You can follow the following steps in this guide, although the dialogues will be quite different.

❖ *Walkthrough*

Start by telling Mattis about the theft. She will show you the entrance to the hideout at the back among the rocks, after going down a ladder. But it's impossible to go through this unless you play druid and transmogrify.

- Go talk to Silfy, who is nearby, he will give you the identity of the leader of the thieves: Mol.

- Then go talk to Doni, the mute Tieffelin child, who is near the oxen and the training area. He will ask you to turn around, before disappearing. You must return to your camp and use a long rest, before returning, in order to make it reappear.

● You can ask him to start again now, which will trigger a perception roll.

● When you succeed, it will reveal a passage that you can take.

❖ Mol and the Tiefling Hideout

Enter the cave, then speak to Mol to ask him to return your belongings. You then have the choice between:

● Pay 40 gold coins

● An intimidation roll, which can have negative consequences later.

● A roll of persuasion.

You can then offer to help her with her next plan, the "Steal the Sacred Idol" quest.

If your stuff wasn't stolen, Mol's attitude will depend on how you treated the children in the Grove. Avoid mistreating them, or allowing them to be mistreated, if you wish to enter into His graces. Otherwise, you will have to succeed at skill rolls.

If you fail, she will report you to the guard, which may cause some unpleasant misunderstandings in town. You will then have to persuade the guards, confront them, or end up in prison.

Selune's Chest

There are quite a few more or less subtle puzzles awaiting you in Baldur's Gate 3, and that of the Temple of Selune is one of the least intuitive. The interface is probably the main culprit in this case. Here are the essential explanations for opening the box.

❖ How to solve the riddle of Selune's chest?

● After entering the Owlbear Cave, take the passage on the left to reach the small hidden temple of Selune. With a statue, a chest, and a book. If you try to open the chest, you will suffer an energy discharge. It cannot be crocheted in a conventional way.

● Start by saving your game, to avoid being blocked if you are unlucky.

● Jump above the water to reach the alcove behind the statue.

● Your characters will then make a Perception Roll. If they all fail, reload the game and start again. If someone succeeds, it will cause a scroll to appear on the ground.

- Left-clicking on the scroll will trigger it to be read, but it's useless. Right-click to select "Pickup".

- Return to the center of the room, then position yourself near the chest.

- Read the parchment you just picked up, while close to the chest. This will unlock it, which will give you experience and an inspiration point if Shadowheart is in the group.

Chantenuit

Events, quests, and strange characters pile up as Baldur's Gate 3 begins, and it's easy to feel lost, or even miss important things. But if it's any reassurance, almost every player feels like they've missed something with Chantenuit, but that's normal To give you a short answer, before elaborating: the rest of the quest only happens in the following chapters of the story.

❖ Find Chantenuit

While visiting the Emerald Grove of the Druids, you will meet a group of adventurers looking for Nightsong, for a certain Wizard of Baldur's Gate. They will mention having lost track of him following the attack on the caravan.

There are quite a few different clues that can lead you on his trail, whether documents, caravan interrogation, or dialogues with other characters. The most notable is the dwarf Brian whose corpse and bag are found in front of the temple occupied by the goblins. It contains a map supposed to lead you to Chantenuit.

❖ Goblin Camp

You must infiltrate the Temple of Selune in the southwest of the map, which is invaded by goblins, in order to find the Selune Outpost in Outland. There are many ways to enter the place, one of the simplest is to free Sazza the goblin at the Druid Grove, then follow her, so that she can testify for you.

Otherwise, you can use different skill rolls, or even the power of the Illithid larva. You can also use stealth and bypass enemies to enter through the hidden passage upstairs, where the goblins are sleeping. Finally, simply killing everyone is also an option.

Once in the temple, your objective is to reach the western section of the area. The easiest way is to speak to the High Priestess of the Goblins, and convince her to speak to you privately. Handle the situation as you wish, then go through the door at the back of the room to reach his private quarters, guarded by an Ogress. You can convince her to let you pass, or kill her.

Then take the corridor which leads to the hidden temple of Selune, with large lunar circles on the ground, which must be rotated.

❖ Hidden Temple: Circular Moon Phase Stones

The solution to the puzzle is simple, and it quickly becomes obvious after a few attempts: You must collect all the dark moons in the light circle. You just have to spin the stones to achieve this, which is not very difficult.

This will open a door on the back wall, which will take you up a very long ladder. Taking it will take you to the famous Selune Outpost in Outland.

❖ Selune Outpost

Explore the area and read the documents present. The only major obstacle is the presence of two statues launching devastating projectiles of light, if you try to exit through the main door, after opening the portcullis.

To deactivate them, you must destroy the Moonstone at the top of the large statue. Unfortunately, it cannot be picked up, even with a pile of crates or a mage hand. Cast a spell, or shoot it with a ranged weapon to break it.

By searching the corpses and reading all the documents present, the quest log will update to indicate that the trail to Nightsong has been lost for the moment.

No other traces of the weapon can be found in Outland at this time. The only solution is to progress further in the story through Malforge, in order to reach Chapter 2 and the Towers of Hautelune. You will have an opportunity to find Chantenuit's trail later. A guide is coming for the following chapters.

❖ Chapter 2: Gauntlet of Shar

Upon reaching the cursed shadowlands, it quickly becomes apparent that Kéthéric Thorm is invincible, and that his protection comes from a mysterious relic: Chantenuit. Everyone wants to get their hands on it, whether it's Ketheric and Balthazar, Jaheira and her Minstrels, and even the goddess Shar. Check out our Gauntlet of Shar guide below to find out how to reach it.

❖ Grisombre

Reaching the end of the Gauntlet of Shar, you will reach Grayshadow, the prison of Nightsong, which is actually a person: an Aasimar, an angel, and the daughter of the goddess Selune. His immortality is forcibly transferred to Kétheric Thorm. Several outcomes are possible depending on the composition of your group and your actions:

- You can let Balthazar of the Cult of the Absolute kidnap Chantenuit and bring her back to Ketheric. If you're planning to ally yourself with the cult and play as a villain, this is probably the choice to make. Obviously, this has catastrophic repercussions for Jaheira and the minstrels. Some members of the group will also not like it at all.

- The second bad option is to have Shadowheart in the party, along with the Night Spear found in the temple. We must let her kill Chantenuit (or encourage her to do so). She will then become Shar's chosen one. Kétheric is vulnerable, but Jaheira and his friends will still die. Shadowheart will descend into darkness to devote himself to Shar, which will also end your romance, if necessary.

- Finally, you can free Chantenuit, or convince Shadowheart to do so. She then finds redemption, and she frees herself from Shar. Chantenuit then leads the charge against Kétheric Thorm and his army, which allows Jaheira to also lead the assault successfully. After the battle, you can ask Chantenuit to accompany you for the rest of the story. She will stay at the camp for a few days, which allows you to unlock a legendary Spear, but also to have additional options during chapter 3.

❖ Chapter 3: Lorroakan

Upon reaching the lower town of Baldur's Gate, you will reach the famous Magic & Spells store, as well as the wizard offering a reward for Nightsong.

Talk to his illusory image on the store floor, then take the portal on the far left. Speak to Lorroakan at the top of his tower, to clarify his intentions. You now have several options:

- If you killed or delivered Chantenuit to Kétheric, you can tell him the truth. Otherwise, it is also possible to deceive him and make him believe that Chantenuit has disappeared. But that doesn't do you much good.

- Confront Lorroakan there and kill him, in order to plunder his goods and his tower.

- Go report him to Chantenuit. Once in her tower, betray her to deliver her to the magician, which will earn you a reward of 5000 Gold, but which will seriously anger some people. Lorroakan will assist you during the final battle with his tower.

- Same as the previous option, but face the wizard with Nightsong to kill Lorroakan. You can freely loot his goods and his tower. Chantenuit will assist you during the final battle, as will Isobelle. If you saved the Tieflings during Chapters 1 and 2, one of them takes Lorroakan's place. He will help you in his place during the final battle. This is by far the best option.

In any case, remember to search the lower floors of the tower, as well as its underground passages. There are plenty of great items, books with permanent bonuses, and more.

Thay's Book of Necromancy

It's both intriguing and terrifying to play with dark powers in Baldur's Gate 3, since the game doesn't spoil the possible consequences of your actions. Here's how to start this series of quests, and what these different branches are.

❖ Pick up the book The Necromancer by Thay

- In order to obtain this book and begin the quest, you must go to the Ruined Village of Act 1. Enter the ruined apothecary shop, to the left of the main entrance. By consulting the register, you will learn that there is a cellar. Open the hatch on the floor and go down.

- Save the game, then approach the crates stacked against the wall, until you trigger Perception Jets. If they fail for all party members, load the game. If successful, a lever will be revealed. By lowering it, a passage will open.

In the next room, you will face some skeletons hidden in coffins, then a mirror will block your way. By having read the books in the cellar of the building, as well as the various letters, you should be able to answer correctly: you are a friend of the master, Szass Tam is an enemy, and herbs are used to heal a wound. Different answers are possible for the final mirror question, some related to classes. Reply "kill the enemies", or threaten to break the mirror next.

Once in the main room, you will see the book, but the door is locked, and there is a trap in front of it. Go to the right of the room, there is a lever to open a passage which takes you back to the entrance to the cellar. And above all, there is the rusty key, which will allow you to open said gate.

- Please note, the book is placed on a pressure plate. If you pick it up, it will trigger traps. It's better to defuse it.

- It is advisable to have Astarion in the group before touching the book, if you wish to take one of the specific branches of the quest.

❖ How to unlock or destroy the ancient book?

By interacting with the book, a dialogue will take place with the members of the group. Right now, you don't have many options, since the book is firmly closed. You only have two options at the moment:

- Destroy the book: Place it somewhere, then use Radiant damage to destroy it (Ex: Shadowheart's Tracer Bolt). This will instantly give you a lot of experience and make several Shadows appear who will engage in combat, which will also give you XP. The quest then ends there.

- Open the book: To do this, you need a Dark Amethyst, to be collected at the bottom of the spiders' nest (see further). Which allows the quest to continue. This also opens up some additional options later.

- Give the book to Astarion: If Astarion is in the group when you pick up the book, he may show interest in it. You can then hand it to him, which triggers special interactions. It is possible to continue the quest all the same, but it is Astarion who will have to read the book.

❖ Obtain Dark Amethyst

If you want to open Thay's book, you must venture into the well in the ruined village, then face the phase spiders found there.

In the second part of the cave, a formidable boss awaits you, the Spectral Spider Matriarch. It's a tough fight, although it's not completely necessary to use discretion.

The Dark Amethyst, required to open the book of necromancy, is located at the bottom, at the very back of the cave. You can spot it easily, with its purple halo.

After picking up the Dark Amethyst, all you have to do is right-click on the book in the inventory to launch a new dialogue, which will allow you to open it.

❖ Gayle and Thay's Necromancy

You can also give the book to Gayle after unlocking it, so that he can absorb its magic, which will have the effect of destroying it, as well as temporarily solving his problem. You don't have to have opened the book to be able to give it to him.

❖ Reading Thay's Necromancy

Once the book is unlocked, you can choose to have it read to anyone you want. But it's better to have a character with high wisdom, like Shadowheart. You must also accumulate magical objects offering bonuses to saving throws and various bonuses, such as a potion of heroism before you start reading it. The inspiration spell is also a great help.

You have 3 successive Wisdom saving throws to succeed in order to turn the pages:

- D10
- D15
- D20

To close the book currently being read, there is a D10 to pass.

Certain classes like Wizard and Occultist have bonuses on their saving throws (D5), the same for Shadowheart with a D10.

If you succeed on all saving throws, you receive experience, as well as the permanent ability to communicate with the dead (reroll every day). There is also a continuation of the quest, later in the story.

But be careful, in case of failure, there are heavy penalties:

- If you fail 1 saving throw, you are afflicted with a penalty to wisdom saving throws for 50 turns.

- If you fail 2 rolls, then you have a permanent penalty, which imposes a disadvantage on wisdom rolls.

❖ How to read the rest of the grimoire?

After having succeeded in all your rolls, you will discover that you are invited to continue reading. But by testing this option, the book resists you, and it is impossible to go further. The solution will not arrive for a long time, a very long time even. Indeed, we have to wait for chapter 3 of the campaign. Once you arrive at Baldur's Gate, you must infiltrate the cellar of the Magic & Spells store, then solve the puzzle of the doors in order to reach the secret tomes found there.

You must read the book "The Triarch Codex". It allows you to read Thay's Necromancy without having to make saving throws, at least if you haven't already, and above all: you will be able to read it to the end. This will unlock a powerful new Necromantic power: summoning a group of ghouls capable of paralyzing your enemies. This is probably the best summoning spell in the game.

Matriarch Phase Spider

The difficulty in Baldur's Gate 3 is not linear, it is possible to very quickly encounter enemies who will make short work of you if you go to the wrong place at the wrong time. The Matriarch Phase Spider is the main example, after the fight aboard the Nautiloid. She can kill your strongest characters in a single turn if things get out of hand. Discover different ways to kill her.

❖ Position of the Matriarch Phase Spider

You can find this boss under the Ruined Village overrun by goblins, from Chapter 1 There are at least two ways to enter the area:

- Go down the well rope, in the middle of the village (Investigation Jet).

- Enter the lower level of the forge, then locate the fragile wall (Perception Roll), you must then demolish it with a blunt weapon or explosives for example.

This will take you to the phase spiders' nest. A first fight awaits you against 2 phase spiders and 2 Ettercaps. After defeating them, you will be able to enter the second part of the nest, with 2 other phase spiders, and above all, the matriarch.

❖ Preparations

Before facing this formidable boss, it is better to prepare for the fight:

- Level 3 is strongly recommended for this fight, level 4 is more appropriate. This is not mandatory using some rather deceitful methods, but it makes things a lot easier.

- Take a long rest to recover all your spells and abilities.

- Distribute the consumables among the group members. Haste potions will work wonders in this fight, as will elemental arrows (fire, acid). Scrolls can also make the difference if the fight drags on.

- Remember to search the area to collect the spider boots which provide immunity to the web. Lae'zel or another melee character will really need it.

- Confronting the 3 Ogres in the village led by Enlightened Lump before going down into the spiders' nest is a good idea. Negotiate with them, and promise them 1000 gold if they fight for you. Their leader will then give you a Foghorn allowing you to summon it. If you use it in the spider room right before starting the fight, it is possible that they will win the fight for you, especially if you intervene. The only problem is that you will potentially lose some experience. Consider finishing them off, or searching their bodies to obtain the Crown of Intelligence.

If you are patient and sneaky, you can also infiltrate the boss room, and assassinate the two phase spiders using stealth and a back attack with Astarion for example. You can also destroy the spider eggs nearby, to prevent the boss from being able to call them for reinforcement. This will make the fight much easier.

❖ Battle Tale Matriarch Phase Spider

There are many ways to take on this boss, depending on your party composition and its spells. But a few methods are particularly effective:

- Use a monster summon or lure to distract the boss.

- Using the Level 1 Thunder Wave spell with Gayle (or other characters) can propel the boss to make it fall off a cliff. This will deal huge damage to him. You can even repeat the operation several times. He can even die instantly if you throw him into one of the big chasms, but picking up your loot afterwards may be difficult.

- Fire arrows and other AoE consumables allow you to kill the bunch of small spiders summoned very easily.

- Likewise, these arrows, area attacks, and multiple attacks like fiery rays make it easy to destroy the cobwebs the boss is on. He's going to fall down and get really hurt. It is even possible that he will lose a turn if he is considered "Down".

- A particularly effective consumable on this boss is the Manticore Poison that Nellie gave you at the Druid camp, if you handled the dialogue well. This boss has 50% resistance to poison, but he will still take huge damage.

- Having characters positioned high up, with a lure located between them (thanks Shadowheart) while the boss is below is a fantastic way to easily hit him. Use abilities like Shadowheart's Tracer Beam to kill him even faster.

- Around 50% life, the boss will become enraged, and he will gain 2 attacks per turn. They are formidable, both at a distance and in melee. Remember to space out the group members, to avoid them all dying in a poison AoE. It's best to save your most powerful spells and abilities for the end of the fight, in order to finish off the boss quickly.

❖ Awards

In addition to the experience, you will obtain a Poisoner's Robe, which can potentially prove useful for a character specializing in poisons.

And above all, you will be able to pick up the Dark Amethyst at the bottom of the area, which will allow you to open Thay's Book of Necromancy.

Investigate Kagha

The Druid Grove is at the heart of the intrigues of the first chapter of Baldur's Gate 3, there are many quests there, and the choices you make will dictate its future. This will influence other characters, and elements in the sequel. Apart from the problem linked to goblins, it is the conflict between Kagha and the Tieflings which will attract your attention.

❖ First meeting in the Emerald Enclave

Upon entering the enclave, you will encounter Kagha threatening a girl, caught while trying to steal the Idol of Silvanus. You have two options during this dialogue with different ways to achieve this:

- Convince Kagha to free Arabella, via your dialogue options or skill rolls. This will please Rath, who is nearby. You can also get a reward from his parents when leaving the building.

- If you do nothing, Arabella will try to escape, and the snake will kill her.

You can then speak to Kagha, who encourages you to speak to Zevlor, to help lead the Tieflings out of the enclave.

If you talk to Kevlor, in the section of the enclave full of Tieflings, he asks you to assassinate Kagha. If you go to talk to Kagha next, you can start the fight, which will involve the nearby druids.

❖ Investigate Kagha

A better way to deal with the rest is to go and loot Kagha's chest, hidden behind a corner, in the room at the back left. In the same building where you can find her. This counts as stealing, but since the wall blocks vision, you shouldn't get caught.

Inside, you will find a document that you will need to read, indicating that Kagha has a secret meeting with shady individuals. You can collect additional documents at the meeting point.

To do this, you must go to the southwest of the map. Cross the woods, then the marsh. You will then have to jump from rock to rock to reach a small island. A fairly difficult fight awaits you there, against treants and mud mephits. Be careful, they explode violently when they die, it is better to move away and kill them from a distance.

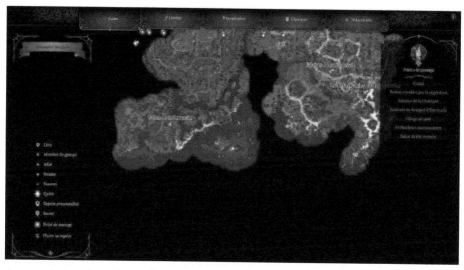

Approach the gigantic tree in the center of the island, there will be a Perception roll to discover the crack in the bark, with the documents. Warning: you cannot discover it without having consulted the documents in Kagha's chest.

You can now return to the Druid Enclave and speak to Kagha, in order to confront her, evidence in hand. The 3 rats in the room will transform, which will reveal shadow druids who will attack you.

You then have two options:

● Attack Kagha and the Shadow Druids, then kill them.

● Make two skill checks while speaking to Kagha, to persuade her to come to her senses. She will then assist you against the Shadow Druids, which will lead to a more positive conclusion.

Note in passing that you can cooperate with Mol at the same time to steal the Idol of Sylvanus, which has other possible repercussions.

❖ Denounce Kagha in Halsin

If you don't confront Kagha, but have the documents, and intend to take them directly to Halsin, the head of the grove, you may be a little disappointed.

After rescuing Halsin from his cage in the goblin camp, and killing their leaders, you will be able to follow the conversation he has with Kagha at the grove. He will then reprimand her severely, and appoint someone else in her place, at the head of the grove. Then speaking to Halsin, and mentioning Kagha's connection with the shadow druids, he will simply respond that the grove needs her. Things remain there, without reward or notable consequences.

stone disk

While exploring the Temple of Selune transformed into a goblin base in Baldur's Gate 3, you will eventually come across an almost deserted area: the quarters of the goblin priestess of the Absolute. By following the corridor, you will arrive in a forgotten temple, with stone discs to rotate. The good news is that they are linked to the Nightsong quest.

❖ Hidden Temple: Circular Moon Phase Stones

The operation of the stone discs is very simple, you just need to make a relay between the 5 moons, rotating them in turn, to bring the black moons to the right place. It is necessary to rotate each outer moon, to bring the black moon to the junction with the central moon, before rotating it in turn. By alternating rotations, you can transfer all dark moons into the same outer moon.

The solution to the puzzle is also not very difficult to find after a few combination attempts. You must collect all the dark moons in the bright circle, lit by the moon. You just have to spin the stones to achieve this, which is not very difficult.

This will open a door on the back wall of the room, which will take you up a very long ladder. Taking it will take you to the famous Selune Outpost in Outland. A lever will also appear on the wall, if you were unable to detect it with Perception.

stone disk

While exploring the Temple of Selune transformed into a goblin base in Baldur's Gate 3, you will eventually come across an almost deserted area: the quarters of the goblin priestess of the Absolute. By following the corridor, you will arrive in a forgotten temple, with stone discs to rotate. The good news is that they are linked to the Nightsong quest.

❖ Hidden Temple: Circular Moon Phase Stones

The operation of the stone discs is very simple, you just need to make a relay between the 5 moons, rotating them in turn, to bring the black moons to the right place. It is necessary to rotate each outer moon, to bring the black moon to the junction with the central moon, before rotating it in turn. By alternating rotations, you can transfer all dark moons into the same outer moon.

The solution to the puzzle is also not very difficult to find after a few combination attempts. You must collect all the dark moons in the bright circle, lit by the moon. You just have to spin the stones to achieve this, which is not very difficult.

This will open a door on the back wall of the room, which will take you up a very long ladder. Taking it will take you to the famous Selune Outpost in Outland. A lever will also appear on the wall, if you were unable to detect it with Perception.

Tatie Ethel

Perhaps inspired by Auntie Danielle, Auntie Ethel is undoubtedly the most deceitful, deceptive and evil character encountered at the start of Baldur's Gate 3. Even Nere is almost sympathetic next to her. But if you manage the situation well, this is one of the rare opportunities to obtain a permanent characteristic bonus.

❖ First meeting at the Druid Grove

You will meet this charming old lady in the Druid Grove, near another old woman distributing food. She will give you a healing potion, while inviting you to share your worries with her. Do not drink this potion, throw it away, it is poisoned.

You have to acknowledge being afflicted by an Illithid parasite to unlock the sequel, even if Shadowheart doesn't approve. She will then arrange to meet you at her tea room, in the sunny marshes.

It also has different consumables that you can buy, if you want.

❖ Getting Auntie Ethel's Help: The Two Brothers

Go to the southwest of the Ruined Village to find auntie. You can either go through the village, at the risk of confronting the goblins, or jump below from the road. You will normally come across an altercation between Auntie Ethel and two men who accuse her of being responsible for the disappearance of their sister, Mayrina. You will make a Perception roll to determine whether you notice that the old woman's attitude is suspicious or not. Three outcomes are possible, and they do not have a great influence on what happens next:

- Joining (rightly) the brothers' cause against Ethel. She will then disappear and reappear in her tea house, at the other end of the marshes. You can then talk to the brothers, and offer to help, but they will refuse and go save their sister, which leads to their death.

- If you do nothing, Ethel blames you for not helping her. She kills the brothers directly before disappearing, as with the previous answer.

- By choosing to defend Ethel, the brothers attack you, and you must kill them. In summary, there is no right choice, but at least the last option will give you experience.

- Unless you're playing Paladin, it's best to stand up for Auntie Ethel, especially if you intend to negotiate with her later.

❖ Sunny marshes and tea room

You must then venture into the green swamps. You will quickly be entitled to a perception roll. If successful, the gigantic illusion that covered the area disappears. It's a horrible, putrid swamp full of corpses, traps and monsters that pretend to be sheep. Don't walk in the water, there are devastating traps there. Also, do not eat apples found in the area.

- As you progress through the area, you will encounter Gandrel, the vampire hunter, and the Tea Room. Inside, you will meet the sister, Mayrina, and Auntie Ethel, who forges her food. If you try to bring up the subject of brothers she will threaten you. If you insist on mentioning their death, she will make Mayrina disappear before revealing her true form and disappearing into the chimney.

- If you avoid the subject of brothers, and mention Removing the Parasite, Auntie Ethel claims to be able to cure you in exchange for one of your eyes. All your companions will disapprove. Obviously, the procedure will fail. You will lose an eye, which will inflict a permanent penalty on you, and the larva is still there. You should not accept his help.

❖ Revealing Ethel's true face

- If you dispel the illusion at Ethel's chimney, or insist on talking about the brothers, she will reveal her monstrous form. She will then turn invisible and flee, while 4 elves will attack you.

- By being well prepared and starting the fight yourself, it is possible to significantly reduce Ethel's life before she manages to escape. It is even theoretically possible to kill her, which will modify the events that follow in her lair. Some traps disappear, and you can save some NPCs. To achieve this, you can talk to her with one character while the other 3 stab her in the back for example.

In any case, go through the door revealed in place of the hearth, and enter its secret lair.

❖ *Tunnel overgrown with vegetation*

Welcome to the Overgrown Tunnel, Auntie Ethel's private dungeon.

- There is nothing you can do to save the unfortunate people who made a pact with Auntie Ethel. Even the poor petrified dwarf is doomed. Some people will get better after killing Auntie.

- It is possible to pass through the closed door, as if it did not exist. Successful perception rolls allow you to obtain the solution.

- Do not equip the terror masks placed on a table. You will be afflicted by a mind control effect. You can nevertheless pick them up to use them in an original way, a little further.

- A fight normally awaits you after the door, against the 4 poor mask wearers (unless Auntie is already dead). It's best to start the fight with a sneak attack. You can also avoid them completely, to go talk to them after the boss dies.

- The next section requires climbing down vines full of trapped flowers and poison clouds. Using Fire Bolt, or another fire effect, will cause the gas to explode and destroy the flowers, but it will not solve the problem. The best way to stop the gas is to throw an object at the vent, such as a crate, pot, shield, mask, etc.

- Once at the bottom of the area, you will see Mayrina in a suspended cage. Auntie Ethel is invisible, she is hidden in the area. Approaching discreetly from the right, then circling the area, is a good way to start the fight much closer. You can also position your characters. Try to place your mages and archers high up.

- Once spotted by Auntie Ethel or Mayrina, the fight against the boss begins.

❖ *Boss fight*

Auntie Ethel is a pretty formidable boss, especially if your level is low, and the group composition is not ideal. She will start by setting fire to Mayrina's suspended cage, before making illusory clones appear capable of inflicting damage on you.

- If you are afraid that the fight will drag on, and that the cage will be destroyed by the flames, you can put it out with a water creation spell (or the staff that grants this spell). This has the advantage of wetting Mayrina, which will prove useful later.

- The Magic Projectile spell is absolutely fantastic in this fight, since it allows you to destroy all illusions at once.

59

- You can find out who the real Auntie Ethel is by inflicting a negative effect on her, or a spell like Witch's Bolt, which will bind her to the caster.

Subsequently, Ethel will teleport Mayrina outside of her cage, before taking on her appearance. You must then attack the right target, otherwise you will kill the poor girl. If Mayrina is wet, or you applied negative effects to Ethel, you can tell them apart easily.

It is possible to kill Auntie Ethel instantly with a bit of luck. By pushing her into the void. This is however not recommended, since you will lose your loot, and the possibility of negotiating.

❖ Spare Auntie Ethel

By reducing the boss's life enough, she will offer you a deal. Her life needs to be very low to do this, so she may not do it if you kill her with a big attack. The triggering of this dialogue seems quite capricious, it is possible that there are links with the previous dialogues, for example by defending Ethel with the two brothers. Save before starting the fight.

She offers you a permanent +1 bonus in any characteristic of your choice if you choose the power, in exchange for her survival. And then she flees with Mayrina. The alternative is to save Mayrina. You can try to negotiate with a good skill roll to get both at once. This is an excellent option, even if it is not necessarily within the reach of paladins.

Leaving Ethel alive is not such a bad choice, even in terms of story, since she will inevitably come back to life and cause you problems again during chapter 3 during the quest to save Vanra . As well enjoy.

❖ Sauver Mayrina

After Ethel dies or runs away, if Mayrina is alive, she will start insulting you. This idiot doesn't realize that she was about to be scammed by Auntie Ethel and her "monkey's paw" help. There is nothing special to do at this point.

- You can use the Communicate with the Dead spell on Ethel's corpse to obtain some interesting information.

- Then search Ethel's laboratory. There they find documents, a magic stick, and above all a special wand: second divorce. By taking the teleportation portal, you will reappear behind the tea room, and Mayrina is there.

Mayrina will complain about having to carry her husband's body home. You then have several options:

- The first is to do nothing, and leave her to her sorrow, or try to comfort her.

- Bring her husband back to life as a zombie with the wand (not to be done as a Paladin). You can then give her the wand, and she announces her invention to restore her identity at Baldur's Gate, which will lead to a new encounter later in the story.

- You can also bring her husband back to life, but keep the wand for yourself. This allows you to summon a zombie for a few turns. This can help in combat. Mayrina still makes her return during chapter 3, but without her husband.

Gandrel

You will encounter many mysterious characters during your adventures in Baldur's Gate 3, and it is very difficult to judge their motivations and the impact they may have on the main story. There are many possible outcomes, speaking with Gandrel.

❖ Where to find Gandrel?

This character is located in the southwest of the map, at the bottom of the swamps, right next to Auntie Ethel's tea room, in a remote alcove on the left. The coordinates are X: -23, Y: 244.

❖ Possible choice

Discussion with Grandel can lead to drastically different results if Astarion is present in your party or not, which can have long-term repercussions.

◦ If Astarion is present

- By choosing to say nothing and not intervene, Astarion will stab Grandel in the eye by surprise, and kill him directly. After a few lines of dialogue with Astarion, it leaves it there.

- You can direct the conversation to extract information from Grandel, about the reasons for his presence, and why he wants Astarion alive. This allows us to obtain some leads on the events taking place at Baldur's Gate, and indirectly, on the actions of Cazzador the vampire.

- You can choose whether or not to reveal that Astarion is present. But if you do, a fight is inevitable with Grandel. He's a pretty tough opponent at low level, but you should easily overcome him at 4 on 1. It's also possible to simply say nothing and leave the area.

- It is also possible to agree to deliver Astarion to him. Choose to reveal that Astarion is in the group, then abandon him, which will lead to his capture and final disappearance. Your companions will disapprove of this choice. Obviously this is a bad choice, unless you're doing it for role play reasons.

◦ If Astarion is absent

- You can reveal to him that Astarion is in your camp. But in this case, after a long rest, Astarion will disappear permanently. You can warn Astarion that Grandel is coming for him, before you rest.

- You can choose not to say anything.

- It is possible to attack him too.

- Finally, it is possible to promise to bring Astarion to deliver it to him.

❖ Booty

If you decide to kill Gandrel, you will obtain the following rewards:

- 60 XP
- Dagger
- Heavy crossbow
- Leather armor
- Healing Potion
- 2 Torches
- Acid Arrow
- Fire Arrow

Wolf Statue Runic

As you venture into the depths of the Emerald Enclave of Baldur's Gate 3, you will come across Nettie, then Halsin's laboratory, and finally the archives, with a wolf statue surrounded by 4 runic tiles, 3 of which are already placed. But the last one is missing, the wolf rune.

❖ How to get the wolf rune?

It is useless to search the entire Emerald Enclave in search of the last rune, since it is in the possession of Rath, the druid who asked you to find Halsin and who was trying to calm Kagha. You can obtain said rune in several ways. The most natural is to go to the goblin base and complete the quest: Save the first druid, by going to free Halsin, then by going to kill the leaders of said goblins. After returning to the enclave, speak to Halsin, who will direct you towards Rath in order to obtain the famous rune.

You can also obtain the rune through less scrupulous methods, such as killing Rath with the goblins. You can then loot it from his corpse. Pickpocketing is normally an option too, but it's not an easy method to use in the room he's in. You will need to use a potion or an invisibility spell.

❖ Open secret room and rewards

By placing the rune in the dedicated notch, the statue will sink into the ground, which will give way to a staircase, with an animation that is rather pleasant to watch.

The enclave's vault contains three notable items: Sorrow, a rare +1 two-handed glaive that allows you to use a druid cantrip. The Robe of Summer, an uncommon item of clothing granting resistance to frost, and Wyvern Poison, a rare weapon coating.

Complete the Master Weapon

Sometimes the seemingly simplest quests actually turn out to be the most difficult in Baldur's Gate 3. The blacksmith in the ravaged village would certainly have struggled to complete his master weapon without your intervention.

❖ Start Quest

You must read the apprentice's diary in the ravaged village. You can find it at the following coordinates X:30 Y: 426, in the blacksmith's house, to the right of the village entrance. The journal is hidden in a crack to the left of the fireplace.

You must then destroy the cobwebs blocking the hole in the floor, in the next room. By jumping into the hole, you will reach the forge. Open the chest in the northeast corner of the room (Perception roll required). You will also need to unlock it. It contains the master weapon blueprint.

❖ Finding susurreau bark

The first serious step is to find an ingredient present in Outland, the dark depths: the underground region. You can reach it in different ways, but the wisest is certainly to follow the quest of Nightsong: go through the Goblin Camp, then solve the Stone Discs riddle. You can also go through the Zentarim hiding place, or even go down into the giant spiders' nest, there is even access via the forge by destroying a wall. This requires facing phase spiders and a big boss, the Matriarch, then jumping into the central hole with a Light Feather effect.

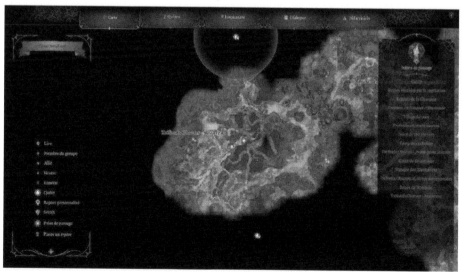

Once in Outland, look for the giant, luminous weapon with anti-magic flowers. It's the susurreau. But be careful, hooked horrors and a crazy dark elf will attack you when you approach.

After eliminating these unwanted people, climb the trunk of the tree until you reach the branches. You can collect the bark from above. Its exact coordinates are

❖ Fusion: Forge the Master Weapon

Bring the susurreau bark back to the forge, but you now need a specific weapon to finalize the weapon: a normal sickle: 1d4 damage. You can find a billhook in the Druid Grove, near one of the druids protecting the Idol of Silvanus, in the middle of the area.

All you have to do is combine the sickle with the forge, to obtain your reward: a Susurreau Sickle +1. It silences the target. This could be an interesting druid or monk weapon early in the game. But it comes way too late in the adventure for that.

Freeing Nere in Malforge

What we can call the end of the first chapter of Baldur's Gate 3 will plunge you into much more difficult and complex situations than before. Crossing the lake into the Outlands, you will arrive at Malforge, in which the central element is the urgent rescue of Nere, an awakened soul of the Absolute. Whether it's to try to infiltrate the cult of the Absolute, or to bring back his head to the patriarch of the mycoids.

❖ How to save Nere?

By advancing a little in the area and speaking to the Duergars, you will learn that Nere is stuck in a tunnel following a collapse, and that he risks dying of asphyxiation soon, as well as the deep gnomes in his company. It appears that this prohibits you from using a Long Rest, otherwise Nere and her companions will be dead when you return. This isn't necessarily a problem, but it does represent a loss of experience, and you may miss out on other gnome- or cult-related opportunities. It's best to take a Long Rest before taking the boat to Grymforge.

To save Nere, the solution is simple: you must destroy the huge landslide that blocks the only access door. Of course, the problem is figuring out how.

❖ Find Philomène's explosives

As you have probably already noticed, the landslide resists even the most violent attacks. By using several powerful explosives dealing force damage, it is possible to destroy the rocks. But apart from that, it is almost impossible to destroy it without using an item found in Grymforge: runic blast powder. By talking to the Duergar, you can learn that a Deep Gnome escaped with a barrel of this precious material, so you must start by finding it.

You can locate it in the northeast corner of the map. There are several ways to reach the area, either by going through the other collapsed tunnel (which can be cleared easily), then jumping onto the suspended platforms, before climbing up to the hall. It's quite a long road. Or, passing through the dining room, and discovering the hidden button at the back, which leads directly to a nearby section. You're going to have to kill a few oozes along the way.

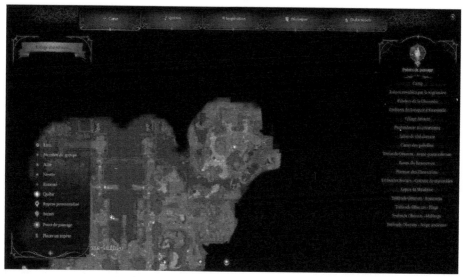

Once there, you will find Philomène and her barrel of explosives, which she threatens to detonate, and you with it. You can convince her to calm down in many ways, from mind reading to persuasion to intimidation to Illithid powers. You can also use a stealth character to steal the barrel behind their back during dialogue, the result is hilarious.

If you ask for powder, she only gives you a vial of explosive rune powder, which is enough to clear the passage. You can also try to confiscate the entire reserve, in order to use it later. But don't use the explosive barrel.

❖ Destroy the (almost) indestructible landslides

Return to the rockfall room. But before destroying them, it's better to decide what you plan to do next. If your goal is to kill Nere and all the Duergar, and save as many gnomes as possible, then it's best to kill the dwarves first. Kill as many Duergars as possible isolated in the area, with sneak attacks, or pushes into the water for example. If you fight everyone at once, the fight is likely to be horribly difficult.

In any case, throw the vial on the landslides. It is advisable to throw it on the top of the rocks. The reason is simple: the explosion is huge, and it is likely to hit the Gnome slaves or the Duergars, which would make them all hostile.

Next, use a projectile or Fire Bolt to detonate the vial, causing a huge explosion. Consider walking away beforehand.

❖ Dialogue with Nere and choices to make

If you were quick, a furious Nere will come out of the tunnel, accompanied by two gnomes, one of which he will immediately execute. You then have several possible choices:

If you want to infiltrate the cult of the Absolute, and extract information from Nere do not intervene to save the slaves. If you decide to protect the slaves, which has positive repercussions for the future, you will have to face Nere and all the remaining Duergars at the same time. Let's clarify that helping Nere and leaving him alive does nothing. This ungrateful bastard will never give you the slightest reward, and he will never help you either.

To make the following fight easier, you can also encourage him to kill the Sergeant. This will make one less formidable enemy. In this case, he will then ask you to kill the Duergars, and he will assist you during the fight. That is, if you haven't killed them before. This is by far the easiest option.

Once the Duergar are dead, a new dialogue begins with Nere, with several possible conclusions. You can choose to kill him and decapitate him to return him to the mycoids. At 1v4 this shouldn't prove difficult.

Consider exploring the second part of the map, with the Adamantium Forge. There are plenty of great items to collect or forge, as well as a big boss.

Arcane Tower

Exploring the Underdark is significantly more difficult than exploring the surface at the start of Baldur's Gate 3, and it's not just about enemies. Even the terrain and puzzles become trickier. The Arcane Tower in particular can get you stuck, if you miss one or two important details.

If you are in Chapter 3 of the game, you may be looking for information on another mage tower, the Tower of Ramazith.

❖ How to enter the Arcane Tower?

The first part is simple, we won't dwell on it too much. You must switch to "Turn-based" mode on the map, and hide your characters behind obstacles before finishing your turn. This way the turrets won't be able to shoot at you.

You can easily destroy the turrets using the level 2 Mage/Wizard/Bard spell: Shatter.

To enter the tower itself, you can either go through the main entrance, or the broken window to its left. Or, you can bypass the remaining turrets, jumping on the giant mushrooms to the left, in order to descend into the garden at the back of the tower. As you will have to go there to unlock the sequel, this is a good option.

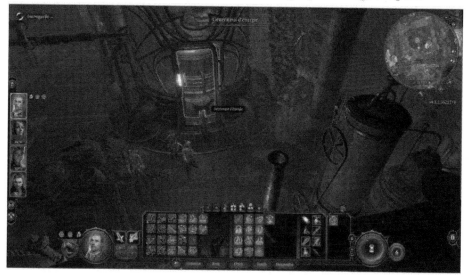

❖ What item should I use in the Energy Generator?

What makes the tower difficult to explore is that it is not powered by power. Magic elevators don't work. The solution is to go down to the lower level, or use the back door if you are in the garden (to lockpick, or to spray with attacks). Which will bring you to a laboratory with the generator.

The open book on the desk and the wilted petals on the ground give you a clue: You must use Susurreau Flowers. They are large blue flowers with an anti-magic aura. You can come across them in several places in the Outland.

Luckily, there's no need to go all the way to the giant tree to get some. There is a small susurreau tree in the garden behind the arcane tower, with two flowers. Right-click, and pick one up. There is no need for both, unless you intend to play without using magic, but with an aura to troll opposing spellcasters.

Click on the generator, then in the empty box, select the Susurreau Flower, before validating.

The tower will then activate, all you have to do is go back to the floor above, to activate the elevators using the "Ascend" and "Descend" buttons.

❖ Secrets of the Arcane Tower

There are plenty of great items to loot in the tower, but a few items particularly catch our attention:

● On one of the balconies, a perception roll will reveal a "Common Chest" containing only mundane objects, such as bowls and forks. This is a second illusion, by picking them up, they will transform into high-value magic items in your inventory.

- The button against the wall, in one of the intermediate floors, requires you to put on the runic dog collar (unearthed from a tomb in the Underdark). You'll just get a steak as a reward, so don't get too tired if you missed it.

- Be careful, on the top floor of the tower, you will face an automaton boss and his colleagues as soon as you exit the teleporter. Prepare for a big fight before you click the button.

- A hidden basement is located below the generator level. You can reach it after killing the boss at the top of the tower. An alternative is to use the Misty Step spell while on the stairs, or Mage's Hand, in order to activate the lever inside and be able to explore it directly.

Forge the Diamonds

Freeing Nere is just the starting point in Malforge, and there are much more profitable things to do in the area. However, this requires a little tinkering occasionally, in order to understand how the adamantite forge available in Baldur's Gate 3 works.

❖ How to access the Adamantium Forge?

You need to go to Malforge in Outland to get started. This involves using the duergar ship to cross the underground lake. This secondary zone is located northwest of the landslide behind which Nere is trapped.

You can easily access the second part of the map using walkways and high platforms, long jumps with high strength characters, a technique of tripling the length of the jump, flight, the Step spell foggy, etc. There is also a fast travel point in the area, you only need one character to unlock it, before teleporting the others.

❖ Find all molds

There are a total of 6 molds to collect in Malforge, you don't have to find them all, but it's better to have more options. You can use the same mold several times. Here are their coordinates on the map of Malforge. You can see the coordinates of the cursor on the mini-map, so you can easily find your way around.

- Scimitar Mold (X: -607 Y: 322) - Adamantium scimitar +1.1d6+1 damage, dazes the target and ignores resistance to slashing damage

- Mace Mold (X: -609 Y: 284) - Adamantium mace +1, 1d6+1 damage, always deals critical hits to objects. Ignores resistance to blunt damage.

- Longsword Mold (X: -625 Y: 410) - Adamantium longsword 1d8+1/1d10+1 damage, dazes the target for 1/2 turns. Always critically hit objects. Ignores resistance to slashing damage.

- Shield Mold (X: -559 Y: 410) - Adamantium shield, +2 AC, dazes an attacker who misses his target for 2 turns, can trigger a reaction to stun the enemy in melee. Makes you immune to critical hits.

- Scale Armor Mold (X: -621 Y: 260): Adamantium scale armor, 16 AC, reduces damage taken by 1, dazes the attacker in melee for 2 turns. Immunity to critical hits. Medium armor, dexterity bonus limited to +2.

- Band Mold: (X -597 Y: 309) - Adamantium Clibanion, 18 AC, reduces damage by 2, dazes the attacker in melee for 3 turns. Immunity to critical hits. Heavy armor, does not give a dexterity bonus.

❖ Obtain Mithral Ore

Adamantium is a magical metal obtained from mithral. Unfortunately, there are very few in the area: 2 in total.

- They can be mined at the following coordinates: (X: -644 Y: 257) and (X: -557 Y: 278), this corresponds to the bottom of the area full of erupting lava and the side passage next to the large staircase leading at the forge.

The ore comes under the forge of large blue rock, which you must break with a blunt weapon. Each time he is in a dangerous area, with lava and enemies, such as the fire elemental, and the lava mephits. Use frost spells, ice arrows and others to kill them easily.

❖ How to use the Adamantium Forge?

Start by saving first, to avoid making a wrong move.

1. Place the Mithral ore in the crucible in the center of the forge. Mithral is consumed after each item crafting. Then use the lever on the edge of the platform to lower it to the bottom of the chasm.

2. Use the mold slot to insert the desired mold. The mold is not consumed after use. You can eject the mold and replace it using the lever next to it.

3. Turn on the lava valve, which will cover all the lower sections of the forge with deadly molten metal. Make sure your characters are well positioned to avoid instant death, or massive damage in the best case scenario. Warning: the first time you open the valve, the Grym boss will enter, he is really strong. Find out how to beat him below.

4. Pull the lever on the edge of the platform, this will bring down the giant forge hammer in the center of the crucible. Massive damage is also dealt to characters and creatures below.

5. Collect your item. You can then change the mold, or press the button on the edge of the platform, to make it rise.

You can repeat all of these operations (minus the boss) once.

❖ How to beat the Grym boss?

This boss has 300 health points and resistance, or even immunity, to almost all types of damage. You also need to be very bad at melee. It's not impossible to fight him normally, but it is long and difficult.

One of the particularities of this boss is that he is made vulnerable by lava, even if it does not directly inflict damage on him. It is therefore necessary to regularly activate the lava valve to fill the room. Of course, this is not without risks, since if your characters walk in it, they are at high risk of dying. Know that you can activate the valve by attacking it from a distance or touching it with a spell, you don't need to stay nearby.

- Another specificity of this boss is that he will target the last person to have attacked him. This means that you can very easily carry him around the entire room, and therefore in the lava, with spells and ranged attacks. Warning: If a melee character triggers an attack of opportunity when the boss moves, he will immediately be targeted.

- The boss is vulnerable to blunt damage when he has walked through lava.

- The boss has two attacks, a melee hit, and an AoE that hits a few meters away. Combine all that with the need to make him walk through lava, and it's really not a good melee fight.

- It's better to place your characters on walkways and ping-pong with ranged weapons. Consider purchasing crossbows and magic bows, as well as using up your ammo.

- The best way to greatly speed up the fight is to use the giant hammer in the center of the room. Try to position your characters to attract him under the crucible in the center, then activate the lever on the edge of the platform. He will take massive damage. You can also attract him with a summon that has his attention. Please note, however, that killing him without using the hammer will unlock a trophy.

Upon death, the boss drops an excellent heavy helmet providing fire resistance, critical hit immunity, and the Hunter's Mark ability.

❖ The best mold to use the Adamantium Clibanion

If you have a character in the party that uses heavy armor, this is definitely the best choice, since it is one of the best armors in the game, and the only very rare quality item (purple) that you can craft at hand. forge. She doesn't really have any competition before the High Moon Towers seller during Chapter 2. You can even make two if you want.

If you're looking for what to craft next, or you don't have a use for this armor, the other items have their charms too.

- Weapons have the advantage of ignoring damage resistances, which is generally great on bosses and tough enemies.

- The other armor and the shield also provide immunity to critical hits, which prevents a lot of unfortunate accidents in combat.

Some players have encountered an issue with lava permanently remaining on the platform. This prevents you from finishing the object and especially from bringing it back up. One way to fix the bug is to switch the movement to turn-based at the bottom right of the interface. After one or two turns of play, it should disappear. You will then be able to go back in real time.

Rugan and missing goods

One of the many charms of Baldur's Gate 3 is the ability to approach each situation in several radically different ways. The quest "Find the Missing Goods" is a good example of this. By managing the quest perfectly, you can get quite a bit of experience and access to unique items from a merchant. But otherwise, it can lead to a bloody fight with chain explosions.

❖ Where to find Rugan?

This quest has three important locations. They are all located to the north of the ravaged village. Take the bridge leading into the northern section of the region. It is better to wait until you are around level 4 to 5, otherwise you will really suffer during fights. Archer Gnolls mean business.

You should quickly find what's left of the caravan, on the road to the east, with hyenas, lots of blood, and a group of gnolls in ambush a little further away.

Going back north, you will encounter a large group of Gnolls, including a boss that you can manipulate with your Illithid larva. They attack a group of guards in the cave, including Rugan.

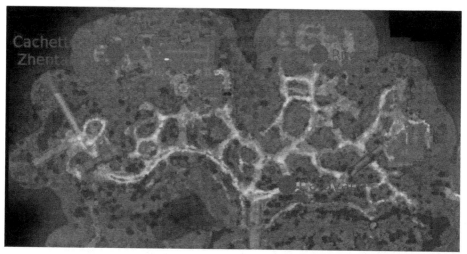

Killing the gnolls isn't too much of a problem, especially if you turn their leader against his troops. Fireballs, spells, or assassinations carried out discreetly can help you well.

❖ Dialogue with Rugan

If everything went well, you can then enter the cave to speak to the leader of the guards: Rugan. Make a backup beforehand, since you need to choose your answers carefully for optimal results. Do not touch the trunk yet.

Here are the answers to give in order to be able to deliver the chest yourself, and obtain better rewards:

● Elturgard is quite far from here. What is your destination ?

Rugan will then announce that he is going to Baldur's Gate to deliver his goods. Then choose the third answer:

● 3 - What exactly are you carrying?

This time, Rugan will respond that he is carrying trinkets in exchange for Tarenths, a currency used by the Zentharim, as the narrator will let you know. Then choose the first option:

● 1 - You are from the Zhentarim. You don't just sell trinkets.

The critical moment has arrived, and Rugan must be convinced to cooperate, with the second answer:

● 2 - [Persuasion] It's your employer's problem, but we can benefit from it. Let's sell the cargo ourselves.

He will then agree to give you the safe, but specifying that it must not be opened. If you decide to ignore his warning, you'll miss out on the best rewards. He will also give you the password to enter the Zhentarim Hideout.

An easier alternative is to kill Rugan and Hiccup (or let them get killed) and take the chest and take it back to the Zhentarim.

All you have to do is pick up the chest with one of your strongest characters, in order to store it in their inventory.

❖ Contents of the trunk

If you're really curious, you can disarm the trap on the chest and then lockpick it. It contains a dead leaf potion and a strange metal vial, containing something. If you open it, it will release a Spectator, a monster that is quite difficult to kill at low levels. Killing him will give you experience, but that's pretty much it. This is clearly not worth it, since the rest of the quest will be considered a failure.

❖ Zhentarim Hideout

As you head west, you will come across Waukine's Rest, a large inn that is burning, or has already burned. It is linked to the quest for the Grand Duke. But that's not what you're here for, move forward into the backyard, with the stables and the storage room (see the map above).

Say the password to the guard at the entrance to avoid a spectacular explosion. You can use Detect Thoughts to learn the password, if Rugan was unable to teach it to you. Then enter the hiding place. A second hidden door is in the corner of the cellar She will lead you to the Zhentarim's lair.

Once inside, Rugan is waiting for you, and you will come across Zarys. The important thing is to deliver the trunk without having opened it. The situation may vary depending on previous dialogue choices, if you have persuaded Rugan to sell the goods, he will be tied up, and you will be asked to kill him, which you may or may not do. If you decide to free him, a fight will begin, which will deprive you of certain opportunities.

If not, you can convince her to let you work with her. You will then be asked to deliver the chest to Baldur's Gate.

You will receive the Heavy Crossbow Hiccup, which inflicts a negative effect on the target if they fail their saving throw. You will also have access to additional goods from their merchant.

You can then decide to massacre everyone and reveal the secret passage to the dark depths, if you feel like it.

Save the Grand Duke

You could say that Larian Studios trolled players a bit with the quest to rescue Grand Duke Ulder Ravenguard in Baldur's Gate 3. It's a bit like Finding Nightsong, but worse. If you're the type who wants to complete absolutely all of your quests before moving on to the next part, it may frustrate you.

❖ Chapter 1: Saving the Grand Duke

You start this quest during the first chapter of the game's story, upon reaching the Waukine's Rest Inn, in the northwest of the map. The inn is on fire, or it has already burned, depending on when you arrive there, but the result is always the same: the Grand Duke has disappeared, he has been kidnapped by the attackers.

You can interrogate or rescue survivors. There are also some documents to read about the corpses inside the burning inn, in order to learn more about their motivations and circumstances.

It is also possible to use the Communicate with the Dead spell on different corpses in the area to extract even more information.

Nothing else can be done in the first chapter, it is impossible to find the Grand Duke at this time. We must move forward in history.

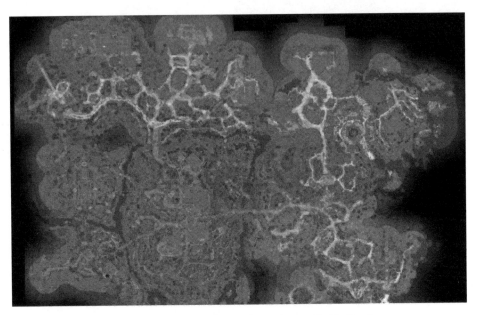

❖ Chapter 2: Where to find the Grand Duke?

Once you pass chapter 2, the quest continues, and you will find other clues concerning the Grand Duke. But it is still impossible to find him at the moment. You must complete the Gauntlet of Shar, and go face Kéthéric Thorm at the top of the Towers of Hautelune. He will then flee, which will open a passage to a previously inaccessible section of the Illithid hive, underground.

By crossing the area, you can free one of the Duke's guards, but he does not know where his lord is. You will eventually catch up with Kéthéric, as well as his two acolytes. The Grand Duke will finally make his first appearance, but the poor man immediately finds himself infected by an Illithid larva, before being pushed into a teleportation portal bound for Baldur's Gate. There is nothing that can be done during this chapter either. Resolve the situation as you see fit, then move on to Chapter 3.

❖ Chapter 3: Coronation of Gortash

Once you arrive at Baldur's Gate, you can speak to the Grand Duke for the first time, during Gortash's coronation at Dracosire Rock, at the end of the large bridge at the entrance to the city. Mizora is at the entrance, and she encourages you to invite Wyll to the coronation. During the coronation, the Grand Duke appears, but you are forced to remain silent, otherwise you will start a fight that you have almost no chance of winning. If you engage in combat, you will have to kill the duke. After the coronation, you can talk to the Grand Duke with Wyll, but nothing special will come of it.

❖ Le pacte de Wyll et l'ultimatum de Mizora

If you saved Mizora at the end of chapter 2, she will appear one night in your camp, during a long rest. Her appearance is rather spectacular, and she will offer Wyll a painful choice:

- Save the Grand Duke in exchange for the eternal damnation of Wyll's soul.

- Put an end to Wyll's pact (he retains his powers), but the death of the Grand Duke is almost assured.

It is indeed a difficult choice. We advise you to end Wyll's pact, at least, if that is what seems most attractive to you from a roleplay point of view. Indeed, as we will see, it is still possible to save the Duke. However, it is not easy.

❖ How can we save both Wyll and the Grand Duke?

Saving Wyll and the Grand Duke at the same time is possible, but you must complete a very specific sequence of quests. You can start by going to the Steel Guard Foundry southwest of Baldur's Gate, and enter the premises by force, or by using stealth. Talk to the Gondian prisoners after eliminating their executioners, and to deactivate the execution devices: they look like purple crystals. The leader of the gnomes will tell you that their families are taken hostage and imprisoned somewhere by the Bainites.

Enter discreetly into the second part of the factory, through the door at the top right. Don't use the freight elevator, you will be spotted instantly. Send a single character, preferably your Dodger or equivalent, with an invisibility spell. Your goal is to check the book at the bottom of the stairs: Submarine Maintenance.

Turn around, talk to the Gondian leader again, then exit the foundry. You can also take a detour to the temple of Umberlie south of the quay, to learn that a priestess was killed by a strange metallic monster. They ask you to punish the person responsible.

Go to the sewers of the lower city, then take the passage leading to the surface, to the northwest of the area, at the following coordinates: X: -163, Y: 838. It is a pipe rising towards "A wall greasy and stinking to climb to the top".

You will arrive in the basement of the Flymm warehouses. The door on the left leads to the surface, and the door on the right leads to the underwater dock with the submarine.

❖ Masserouge and the submarine

Only one NPC can be found in the submarine's dock, its designer, Masserouge. Don't kill him! Talk to him to convince him of the way you prefer, to take you to the underwater prison. The method doesn't really matter. It is strongly advised to take a long rest and optimize your group before entering the submarine. Consider taking consumables, like elixirs and potions of haste.

As you approach the underwater ruins of the Iron Throne, you will enter into magical communication with Lord Gortash, who is not very happy to see you disembark. You have two choices:

- Turn around and abandon rescuing prisoners.

- You dock at Iron Throne, but Gortash will blow up the prison. You will only have a few turns to evacuate as many prisoners as possible.

We didn't try to kill Gortash before going to free the prisoners, since that would be incredibly difficult. Or it would force you to destroy the Steelguard Foundry before freeing the prisoners. There is no perfect solution.

❖ *Evacuate all prisoners from the underwater prison*

We might as well say it clearly, this mission is extremely difficult. Even with excellent preparations and playing well, achieving a perfect result is not cheap. We advise you to make a backup before getting into the submarine and prepare as many items as possible to make your life much easier:

- Haste Potions (use when there are only 3 turns left)

- Items or spells with the following effects: Misty Step, Dimensional Portal, Swiftness, Haste, Heel Click, Freedom of Movement

- Also remember to cast the Great Stride ritual spell on the entire group while you are in the submarine. Also summon as many creatures as possible as reinforcements, such as elemental Myrmidons, Devas, etc.

Cast your limited-time spells right before exiting the submarine, like Haste or Blessing. Their duration is 10 turns, you will only have 6 turns to evacuate everyone and return to the submarine (there is no turn 0). Combat begins automatically by entering the submarine's airlock.

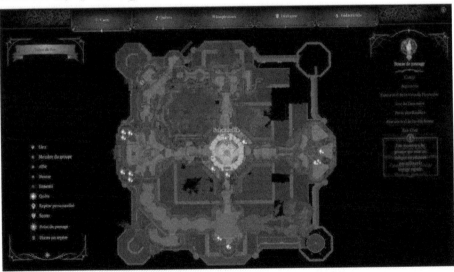

You must climb down the ladder, then split up your group to free the prisoners. There are 4 corridors available, knowing that one of them does not contain prisoners. Prisoners are indicated by markers on the map, you cannot miss them. There are opulent chests present in the more remote passages of the area, but they contain nothing worth the effort of opening them: potions, elixirs, scrolls and gold. This means you only need to send your characters into 3 corridors only, ignore the one without prisoners.

The area is full of sea monsters, the Sahuagis. They are quite painful since they can kill prisoners after their release, and above all, they can throw nets capable of immobilizing a target for 2 turns. You can free it by dealing fire or slashing damage Use your summons as a priority to kill the Sahiagis, since they cannot free the prisoners.

Use a movement ability like Misty Step, and the "Dash" action with all your characters to maximize their movement distance.

You must free all the prisoners in the second round (4 to 5 rounds remaining), otherwise they risk not evacuating in time. This involves sending at least 2 characters into the hallway full of black powder. The cells to be opened via the levers are very far apart, and a single character may not be able to cover the entire distance unless they have Haste.

Of course, the objective in this guide is to release the eagle owl as a priority. It is located in the central cell of one of the passages. It's relatively easy to get to, but complications await. It is advisable to send him 2 members of the group, including a healer and summons to rescue him.

❖ Keeping the eagle owl alive

Unlike other prisoners who are allies, the Grand Duke comes directly under your control upon his release. Heal him as much as possible with your characters. If you don't have any good healing spells, throw high level healing potions in his face, it's brutal, but it works well. A potion of fire resistance might also be a good idea. Also prepare your characters by positioning them a little in front of the cell, even after having freed it.

When it's the Grand Duke's turn, choose the "Run" action then exit the cell. If you chose to sacrifice the Duke, during the dialogue with Mizora (see above), this bitch will appear. She will force the Grand Duke to kneel, before making 4 explosive spiders appear nearby. Managing to kill them is not enough, because their explosion will hit the duke anyway. If you followed our advice, and brought enough summons and characters, you should be able to kill the enemies and evacuate the duke. It is important to use the "Help" action on the Duke with one of your characters, this will end the kneeling, and he will immediately regain his turn.

All that remains is to use the "Dash" action once more, and start running towards the exit. You can use the Dimensional Portal to teleport one of your characters with the Duke, it's a huge time saver.

❖ Return to the surface

All of your characters must be on board the submarine before the end of the last turn. It seems that having a character in the intermediate airlock between the two ladders is still considered saved. This avoids leaving with the submarine before the prisoners have had time to take their turn.

❖ The secret of the dragon

A dialogue with the Grand Duke will begin in the submarine, and another back at the camp. You will be able to speak to him, in the company of Wyll, in order to lead to reconciliation, and the revelation of a great secret: a Bronze dragon is under the city, and he can help you during the crisis, but it is another story.

In any case, congratulations, you have finally saved the Grand Duke, he will be a powerful ally during the last act of the game. You can also go talk to Mizora to make fun of her! But she obviously hasn't said her last word.

Philomène's hiding place

While Baldur's Gate 3 generally offers immense freedom of action to progress, this is not really the case in the quest: Free Nere. At least, to clear the rocks blocking the way. It is almost impossible to destroy them without the rune powder held by Philomena.

❖ Where to find Philomène?

As you have probably already noticed, the landslide resists even the most violent attacks. It is in practice impossible to destroy it without using an object found in Malforge: runic explosion powder. By talking to the Duergar, you can learn that a Deep Gnome named Philomena escaped with a barrel of this precious material, so you must start by finding it.

You can locate it in the northeast corner of the map. There are several ways to reach the place, whether by going through the other collapsed tunnel to the northwest (which can be cleared easily with the help of the animals), then jumping onto the walkways and suspended platforms, before climbing up to the hall. It's quite a long road. Or, passing through the dining room, and discovering the hidden button at the back, which leads directly to a nearby section. You're going to have to kill a few oozes along the way.

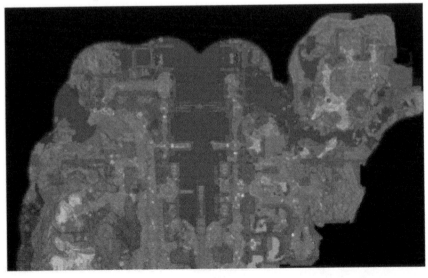

Once there, you will come across a locked door. You can simply go around it to the left, a wall has collapsed. Then, you will find Philomène and her barrel of explosives, which she threatens to detonate, and you with it, if the discussion goes badly.

❖ Dialogue with Philomène

You can convince her to calm down in many ways, from mind reading to persuasion to intimidation to Illithid powers.

If you ask for powder, she only gives you a vial of explosive rune powder, which is enough to clear the passage. You can also try to confiscate the entire reserve, in order to use it later. But don't use the explosive barrel on the rocks, unless you want to kill everyone within 60 feet of the center of the explosion. You can use it to instantly wipe out any group of enemies in the game, if you don't get blown up with it.

An original and hilarious method of managing the dialogue with Philomène consists of sending a character to pick her pockets during the dialogue, in order to grab the vial. You can also take the opportunity to steal the barrel of runic powder. The poor thing will only have her eyes left to cry with.

Chipped stone

After defeating Commander Zhalk (or not) aboard the Nautiloid, and crashing on the beach with the Nautiloid, you can start exploring the first chapter of Baldur's Gate 3. The area is full of quests, enemies, of teammates like Gayle and Karlach, but also of secrets and enormous ones.

❖ How to reach the chipped stone?

As is often the case in this game, you will have to pay attention to the environment and jump to the right place. A small side passage is in the same area as Astarion. You must make a jump below, in order to get down from the rocks.

Then make a save before moving forward into the area. Indeed, unless you succeed in a Perception roll, the famous chipped stone will not appear. Consider using the Assist spell on your character with the best Perception score, such as Shadowheart or Astarion.

❖ What to do with the Chipped Stone?

Once the chipped stone is revealed, the procedure to follow is not intuitive. At least if you haven't played Divinity Origin Sin 1 or 2. In fact, you can move objects in the environment with the cursor, by holding down the button. The principle is exactly the same as moving a file from one folder to another on Windows. You can also right-click and choose the "Launch" option. There is however an important prerequisite, the character you control must have sufficient strength for this. There is no need for the Shovel, however, although it is useful in other circumstances.

Ideally, you should select Lae'Zel and take control of it, or Karlach. They will be able to move the chipped stone easily. This will reveal a box. If these characters are not in your group, click on the return to camp icon to change the group composition.

❖ Awards

The chest contains speed potions, Harper's notebook, which will launch a quest related to the Harpers, and some miscellaneous objects, such as a harp, gold and a ruby.

Omelius

Baldur's Gate 3 reestablishes many races often perceived as monstrous and aggressive, with civilized, even friendly, representatives. Blurg and Omeluum are good examples, with an interesting quest, which will earn you some nice rewards.

❖ Where to find Omeluum?

This NPC is not directly present on the map. To begin, you must go to Outland, the dark depths. Then enter the Myconid village, convincing their monarch that you are not a threat. You must then speak to Blurg, the red Hobgoblin, and explain your problem to him, so that he can call an expert.

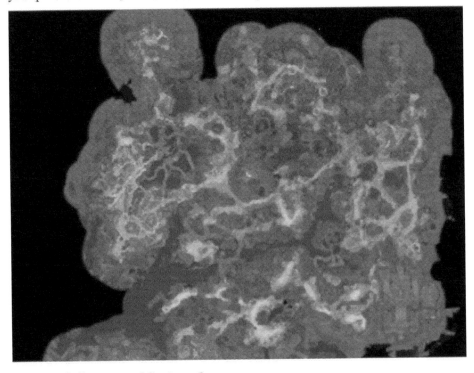

❖ Dialogue with Omeluum

The expert is a Mind Flayer, which makes sense. But he managed to free himself from the control of the master brain. In order to investigate the parasite, he asks you to open your mind to him. To progress in the quest, you must accept.

- The rest of the dialogue is rather obvious, agree to collaborate and search for the components required for the potion: mushrooms.

- You can also trade with him. He sells an excellent Amulet of Misty Steps. This is a fantastic item for any character who doesn't have this spell.

❖ Getting the mushrooms for Omeluum

You can find said mushrooms in the Arcane Tower, also located in Outland, in the southwest corner. Collect the susurreau flowers in the garden, to put them in the generator in the cellar. As you climb the tower, you will find mushrooms in a flower bed. Harvest them, then bring them back to Omeluum.

❖ Drink the potion

Take the mushrooms back to Omeluum, then agree to drink the potion.

Make a save, as there are a few saving throws ahead of you. Missing them can have negative consequences, although not all effects have been well documented. They seem to be more contextual than anything else.

- Inspect the available saving throws each time, in order to choose the one in which you have the best bonuses.

- You will unlock a new unique Illithid power, allowing you to heal a character receiving lethal damage. However, it is quite difficult to use unless you play a healer, or unlock the possibility of using your Mind Flayer powers, in the form of a bonus action.

- In all cases, Omeluum fails to extract the larva. He gives you a Ring of Mental Barrier instead, providing advantage on saving throws against charm and mind control. However, he doesn't want to give it to you for free.

- You can talk to him about the Nautiloid, make an Intimidate roll, or as a Bard, a Performance roll, so that he agrees to give you the ring.

❖ Chapter 3: Saving Omeluum

This may come as a surprise to players who have already completed the game once, but Blurg can be found in Baldur's Gate. In the Scholars' Lodge, but on the only condition that you didn't sell the Githyanki Egg from the Manger to the scholar in Mountain Pass.

He will ask you to save his friend, held in the underwater prison of the Iron Throne You must use the submarine to reach the area. He is locked in the farthest section of the area, making him quite difficult to save. Fortunately, he has a portal spell, to help you evacuate.

Abdirak

The goblin base in the ruined Temple of Selune is rich in diverse and varied opportunities, but often quite dubious. Abdirak is a good example of this, as he proves to you that what doesn't kill you, makes you stronger. It grants one of the rare permanent bonuses in Baldur's Gate 3.

❖ *Where to find Abdirak?*

This NPC is in the second room on the right, upon entering the goblin base. Between Volo prison, and the tortured prisoner. He will not normally participate in combat if you attack the goblins, but he will leave the area after the death of their leaders. So it's better to talk to him as soon as possible.

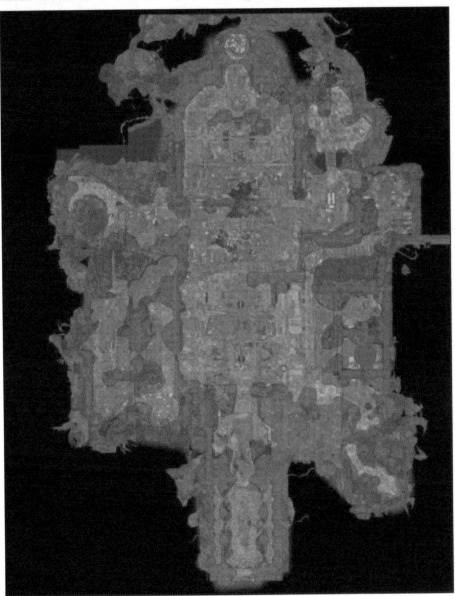

❖ *Dialogue and choices to make*

It is advisable to have Shadowheart and Astarion in your party before talking to

82

Abdirak and starting the torture session, since they will approve of your actions. Conversely, Karlach will disapprove.

You can unlock some additional dialogue options by sneaking into the room to read the book on the cult of Loviatar.

Abdirak must be questioned about his presence in these places and his faith, and ask to be tortured. You can also ask about the prisoner next door.

When asked what you think about inflicting pain without a specific purpose, answer:

● 4 - I thought a loyal Loviatar like you would be in favor of pain.

He then offers to torture you, choose answer 1 to accept.

Abdirak will whip you 3 times, with different possible responses. To get the bonus reward, you have to shout. So scream as much as possible, the Bard even has special Performance rolls for the occasion. You also have to survive the blows, avoid being tortured with a character who is missing life points.

❖ Awards

If you take all three hits, while taking care to scream, you will receive the "Love of Loviatar" bonus granting +2 to attack rolls and saving throws when you are below 30% life.

This is a permanent bonus, but can be lost under certain circumstances, such as being pushed into the void before being brought back to life.

Abdirak has a friendly mace, dealing Necrotic damage. You can steal it from him, or kill him to get it back. There is also a special ax and dagger. This character does not return later in the story, to our knowledge.

Outland

It's always a fascinating moment to discover a huge, somewhat hidden underground region in a game, whether it's Skyrim, Elden Ring, and now, Baldur's Gate 3. You shouldn't be disappointed.

❖ Is it better to go through the Mountain Pass or through Outland?

At the end of Chapter 1, Halsin and other characters will explain to you that the area between you and the High Moon Towers is cursed. Crossing it on the surface would be far too dangerous according to them, and it would be better to go through the depths.

In reality, this is a false dichotomy. The correct answer is "Why not Both?" to use a popular meme. It is strongly encouraged to explore the entirety of Outland during Chapter 1, and to stop at the transition to Highmoon Towers. You can then switch to the side you prefer. We then advise you to go through the Mountain Pass to pick up the legendary Blood of Lathander weapon, which will help you immensely throughout Act 2.

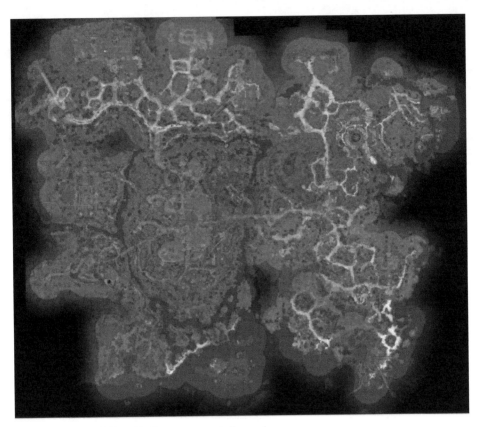

❖ List of Dark Deeps entries

There are so many ways to reach the Underdark that you almost have to do it on purpose to avoid discovering at least one by accident. However, some are better hidden than others.

- The first opportunity is in the ravaged village in the middle of the map. You can jump into the well, or go down into the forge from the Complete Master Weapon quest, before destroying the wall. Both will lead you to the nest of the Phase Spider Matriarch, a fearsome boss at low levels. Using a Dead Leaf potion or spell, you can jump into the central hole in the area, which will take you to Outland. It is nevertheless very dangerous.

- You can go to the north of the map, and do the Rugan quest, in order to enter the Zhentarim's lair. At the bottom of their lair is an illusory wall hiding an elevator to the dark depths.

- Theoretically, it is possible to go through the very well hidden teleporter in Auntie Ethel's cave, but reaching it is excessively complex.

- The best way to reach the Underdark is to enter the Goblin base, for example freeing Sazza the Goblin, then infiltrating the High Priestess's quarters, in order to reach the Shattered Sanctuary and the riddle of the Stone Disks. This will take you directly to the Selune Outpost in the Deep.

❖ Where to go next?

There are tons of things to do in the Dark Deeps, like seeing Omeluum and then the Arcane Tower. But if your goal is to move forward, head to the lakeside, with the Duergar patrol and the docked ships.

Use the ships to cross the lake and reach Malforge, and its Adamantium Forge. The passage to the surface is there.

Baelen

There's a whole new region awaiting you in the Dark Depths of Baldur's Gate 3, and not all local problems are solved by killing everything you come across. For once, it's even the opposite.

❖ Start quest: Find the mushroom picker

This quest can be started in two different ways, either by speaking to the dwarf from the Myconid village named Derryth, or by directly meeting Baelen the mushroom picker, when he finds himself in an uncomfortable position.

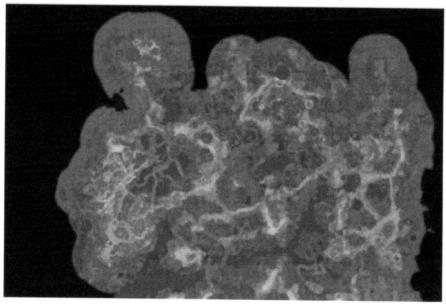

❖ Where to find the mushroom picker?

Baelen is located in a landlocked area, in the northwestmost corner of the Dark Deeps, among explosive mushrooms. To reach it, you normally have to go through the giant susurreau tree a little to the south. Be careful, a fight awaits you there, but this will be an opportunity to discover some secrets linked to the Adamantium Forge and the bark for the master weapon.

❖ Comment sauver Baelen?

As is often the case in Baldur's Gate 3, many methods are at your disposal, depending on the objects and spells at your disposal, but also on your creativity.

The important thing is not to trigger the emanation of spores from the mushrooms, by approaching them or destroying them. The poison cloud is of course toxic, but the real problem is that it will also activate nearby mushrooms. This will trigger a chain reaction that will cover the entire area. A lit torch is also on the ground, at the bottom of the area, which will immediately set the spores on fire and trigger explosions that will kill Baelen, in addition to destroying a useful object hidden in the area.

If you have a scroll of Misty Steps, you can throw it directly at Baelen (Strength throw) or give it to him, if you are next to him. Otherwise, you have to reach for your backpack which contains one. By handing the scroll to Baelen, he will simply teleport out of the area and return to the village.

The secret to crossing the area without triggering the mushroom spores is to separate a member of your group, and switch the movements to turn-based mode. You must then use the "Dash" skill, as well as other techniques such as Misty Step, in order to quickly cross the area without triggering the mushrooms. As long as you don't end your turn near a mushroom, it won't emit spores. In real time, it's almost impossible to pass through the area without triggering anything, whereas with this method it's almost easy. You can also use the Mage Hand spell.

If you recruited Sovereign Glut in Myconid Village, he can accompany you to the area. It does not trigger mushrooms to explode. He can easily throw the backpack and the hidden treasure in the area, the Noble's Foot.

❖ Noble's foot

At the far right of the area, there is a Noble's Foot, a rare mushroom capable of healing almost any affliction, in addition to healing health points. It is destroyed by the explosion of mushrooms in the area (but not by toxic spores). You can use the methods listed above to recover it.

You can then decide what to do with it:

- You can keep it and use it when needed, it is one of the best consumables in the game.

- It can be delivered to Derryth the Dwarf in Myconid Village. This will allow her to open a shop at Baldur's Gate, where she will sell other Noble Feet, which can then be used in any way you like. She will also give you Kushigo's gloves, increasing throws by 1d4 damage.

- If given to Baelen, his mental affliction is cured, but he is discovered to be not a good person. There is no reward.

- By giving it to Shadowheart, she finds some of her memories, including her childhood friend.

Tunnel overgrown with vegetation

Not all enemies are directly displayed as such in Baldur's Gate 3. Evil and manipulative creatures are constantly trying to scam you, such as Raphael and Auntie Ethel. You can find out all the details about the latter in our dedicated guide Here, we will focus on its lair, the tunnel overgrown with vegetation, and the dangers present.

Auntie Ethel Baldur's Gate 3: Saving Mayrina, choices to make and how to beat the boss?

The little old lady from the Druid Grove hides many secrets in BG3. It's an evil monster, a swamp hag, residing in the Tea Room and who will constantly set traps for you. Discover each step, and how to get great rewards without getting scammed in this guide.

❖ How to enter the overgrown tunnel?

Approach the fireplace in Auntie Ethel's house, you must make a Perception check to reveal the nature of the illusion. Engaging in combat with Ethel in one way or another will cause her to flee, which will also reveal the passage. Here is the position of the entrance on the map of the region.

❖ Tunnel overgrown with vegetation: walkthrough

There is nothing you can do to save the unfortunate people who made a pact with Auntie Ethel. Even the poor petrified dwarf is doomed. Some will get better after Auntie Ethel's death, so don't waste your time in the zone, other than to appreciate Auntie Ethel's cruelty the way she manipulated these poor souls.

The first real obstacle is the closed face-shaped door. This is actually just an illusion. It is possible to pass through the closed door, as if it did not exist. Successful Perception and Arcane rolls provide the solution. Save the game, then use Gayle to inspect the door.

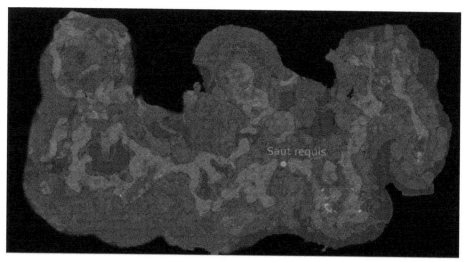

❖ Whispering Mask

The first real trap in the area is on the right, just before the illusory door: 4

whispering masks placed on a table. There is absolutely no need to equip them. You will be afflicted by a mind control effect. This will turn you into Auntie Ethel's slave. If the whole group teams them, you risk Game Over. If only one character is controlled, you can kill them then remove the mask. You can nevertheless pick them up to use them in an original way, a little further. You can also use one of these masks on a character with high Wisdom, like Shadowheart, then talk to the illusory door. This is a way to open it without making the required Arcane rolls. Switching the game to turn-based mode can make things easier without having to make too many wisdom checks.

❖ The 4 masks

After the door, a fight normally awaits you after the door, with 4 poor mask wearers (unless Auntie is already dead). It's best to start the fight with a sneak attack. You can also avoid them completely by using stealth or invisibility, to talk to them after the boss dies.

❖ Poison clouds and exploding flowers

The next section requires climbing down vines full of trapped flowers and poison clouds. Using Fire Bolt, or another fire effect, will cause the gas to explode and destroy the flowers, but it will not solve the problem. You can also destroy flowers with Eldritch Blast and fairly powerful attacks.

To stop the gas, you need to make a Perception check to detect the vent. Rather than rushing into the gas to try to defuse it like a trap, the best way is to throw an object at it, such as a crate, a pot, a shield, a mask, etc. Do not hesitate to come back to Ethel's house to take items and furniture.

By progressing slowly and methodically through the area, you can destroy all the flowers and block all the grates. This way you won't take any damage. Of course, you can also rush into the pile after exploding the flowers, before healing your characters with a Quick Rest. It is also advisable to take a long rest before reaching the end of the tunnel overgrown with vegetation, since a tough fight against a formidable boss awaits you.

Once at the bottom of the area, you will see Mayrina in a suspended cage. Auntie Ethel is invisible, she is hidden in the area. Approaching discreetly from the right, then circling the area, is a good way to start the fight much closer. You can also position your characters. Try to place your mages and archers high up.

Once spotted by Auntie Ethel or Mayrina, the fight against the boss begins.

Sacred idol of Silvanus

The Druid Grove is the focus of the quests in Chapter 1 of Baldur's Gate 3, and there are many ways to handle the situation. We will present to you here what can be considered as one of the optimal ways to obtain the best rewards.

❖ How to get the quest to steal the idol of Silvanus?

Many players missed this quest, including us during our first game. Indeed, you must complete the quest Find your possessions, and speak to Mol in the children's hiding place, but without having carried out any action likely to anger her. This means treating children well, even defending them, whenever possible in the camp.

❖ Steal the sacred idol

Talk to Mol in the children's hideout, and select the answer "I thought you might need some help." She will then explain her plan to you, which consists of stealing the sacred statuette of the Druids in order to interrupt the Druid ritual, and incidentally, reselling it.

Unfortunately, as the ritual is underway, many druids are looking directly at the idol. Even if you manage to steal it, the druids will go berserk, and they will attack absolutely all the foreigners present in the enclave: the Tieflings and your group. Unless your plan is to recruit Minthara after slaughtering the grove's occupants, this is not a desirable outcome.

Kagha must be investigated first. You must either kill Kagha and her shadow druids after gathering the documents, or convince her to turn against them. Once this is done, the theft of the idol will no longer have such drastic consequences.

Once Kagha's account is settled, stealing the idol is not particularly difficult. The druids stop staring at her. Use Astarion or another stealthy character to approach him in stealth mode (c). You can also use spells and/or potions like Mist Cloth, Darkness or Invisibility. The invisibility potion is your friend.

After grabbing the idol, flee the area as quickly as possible. For example, returning to camp. If you are questioned by a druid, you will need to succeed at a persuasion or deception roll. You can then bring it back to Mol, but there is one more factor to take into account: the possibility of keeping it, or recovering it afterwards.

❖ Awards

You must return the Idol of Silvanus to Mol before killing the 3 goblin leaders. Afterwards, she no longer really needs it and no longer gives rewards.

The Idol grants an aura to all your characters, as long as it is in the inventory: +1 to Nature rolls and +1 to taming rolls. As it is not very heavy, it is good to keep it with you throughout the adventure in order to improve your rolls in these areas.

In exchange for the idol, Mol will give you one of the best rings in the game: the Ring of Protection +1 (+1 armor and +1 to saving throws). This is the only way to get this ring, you cannot steal it from him, nor collect it from his corpse.

You can buy the idol back from Mol afterwards, or you can pick his pocket to get it back. Don't deprive yourself of it.

Iron flask

Players accustomed to CPRGs like Baldur's Gate 3 have often gotten into the habit of recklessly searching and opening everything, counting on the fact that this is the best approach to discovering the secrets of the game and obtaining the optimal solution to quests. This is generally the case, but there are a few exceptions, such as the Caravan Chest and the iron flask it contains. You may have to load a previous save, if you regret your choices after reading the information below.

❖ The Missing Goods and Rugan

This quest takes place in the northern part of the chapter 1 map, you have probably already discovered it at this point. Two men defend a chest in a cave against a horde of gnolls. There are many ways to resolve the situation, you can save them, let them die, or kill them yourself. This may lead to different outcomes later. It is better to keep Rugan alive and talk with him, and convince him to give you the caravan chest to take back to the Zentharim hideout, near the Waukine inn.

It is strongly recommended not to open the trunk. It must be picked up and carried to its destination in the Zentharim hideout. If you try to deliver it afterwards, the situation will degenerate upon receipt, since the members of the Zentharim will realize that you have touched the content, and a fight will begin. The problem is this will prevent you from accessing their special wares, including fantastic gloves for thieves and a good magic crossbow.

Of course, it is more than likely that you have already opened the chest, and that your character already has the iron flask in hand, as you read these lines!

❖ Contents of the Iron Flask

Once in possession of the Iron Flask, you have three options: open it, or not open it or give it to Gayle at the camp, so that he can absorb its magic.

If you open it, its occupant will be revealed: a fearsome Spectator aka a Beholder/Tyranneye. A horrible flying monster full of tentacle eyes. Combat begins immediately and your nearby characters will be surprised. This can be a tough fight especially at low levels. He has several attacks per turn, with rays with various and varied effects. You can face another one in Outland, in the small area next to Selune Shrine, if you want to get an idea of   what to expect.

❖ Spectator

Killing this Tyranteye boss grants quite a bit of experience (100-150 XP depending on difficulty), but it's not something you really need. The loot is particularly disappointing, since it only gives alchemy components.

You can use it in a creative way, by releasing it into an enemy camp, while being invisible for example. But the outcome tends to be quite unpredictable, depending on the enemies chosen.

If you want to make the most of the situation, it's better to deliver the goods to the Zentharim, access their store, make your purchases, kill them, collect the chest (or open it, if they left it for you), then then do whatever you want with the Iron Flask. You can also deliver him to Baldur's Gate for a small reward.

CHAPTER 2

Start of chapter 2

It takes about a hundred hours to complete Baldur's Gate 3 for the first time. But it can also be longer, if you're the type to search every crate in the game and inspect every square inch of the map. If you feel targeted, then it is even more important to know when to turn back in each chapter. Otherwise, you risk missing important events, or even completely missing part of the map.

❖ *Warnings welcome*

If it's any reassurance, in the vast majority of cases, Baldur's Gate 3 will warn you when you're about to progress the story past a point of no return. In the majority of cases, this will be an alert message, when clicking on a zone transition. But they are not always found where we expect them.

❖ *Introduction & Tutorial: the nautiloid*

For some players, their character creation is almost as long as a chapter of the game, but other than that, the introduction is rather short. It has even been shortened compared to its early access version. It ends when fighting in the main cabin of the ship, when you activate the main console to try to get it out of the underworld and land it. You normally have enough time to kill Commander Zhalk, who is in the same room.

❖ *Chapter 1*

This first chapter begins on the beach after the crash, but it can end in several places, and in several ways. It is possible to end this chapter very quickly by heading directly to the north of the map, in order to reach one of the two passages leading to the Mountain Pass.

● There is a passage all the way to the Es, continuing straight ahead, after passing the goblin camp.

● The second is to the Northwest, it involves passing Waukine's Rest, the burning inn, and coming across the Githyanki patrol, with their red dragon. If Lae'Zel is in the group, she will probably comment on it.

91

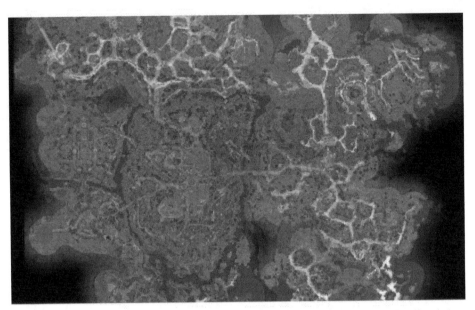

It is recommended to be at least level 4 or 5 to use these passages. Being level 6 or higher isn't a bad idea. They both take you to the Mountain Pass region.

A third passage to chapter 2 is available, it involves going through the dark depths, aka Outland. It is much longer, and you are less likely to take it by mistake. Indeed, you must come across the Duergar patrol, then use their ship to the west of the map in order to reach Malforge.

Once in Malforge, you must take the staircase that goes back to the surface. The nearby Duergar will warn you that this is a bad idea.

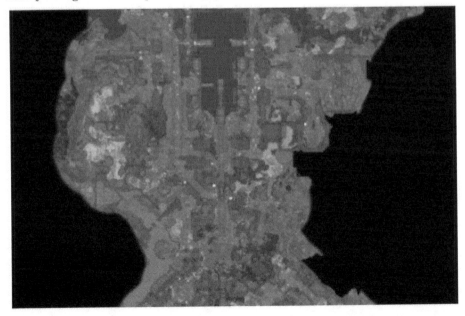

❖ What happens if you move to Chapter 2?

In addition to obtaining a Steam achievement or equivalent, you will know that you are in chapter 2 by coming across a certain magician dressed in red, with a white beard and a pointy hat: Elminster.

We do not claim to know all the events taking place in the event of premature passage to chapter 2. But to summarize: time moves forward, and history with it. Many quests are no longer available, or you can no longer solve them satisfactorily.

The central plot of Chapter 1 is also going to be resolved one way or another in your absence. Different results are possible depending on the actions taken before changing chapters, but by default, Tiefling refugees are massacred by goblins. They will therefore be absent from the rest of the story, which has many ramifications. Many quests and characters in subsequent chapters will be absent, or different.

If it's any reassurance, it's still possible to go back and explore the maps in Chapter 1, even after moving on to Chapter 2. Some quests, like Complete the Master Weapon, are still available.

They sang Lathandre

Welcome to Chapter 2 of Baldur's Gate 3, after very long hours in Chapter 1. To celebrate the occasion, here is a ruined monastery, with puzzles a little more complex to solve than the Selune Chest or the Tower arcane.

By climbing almost to the top of the monastery, you will arrive in a circular room with a large stained glass window in the center, and 4 altars around it.

❖ Find the weapons to place on the altars

The developers are not too cruel, one of the altars is already active: the weapon placed on it, a long sword is illuminated, just like the two crystals at the foot of the altar. The clue is clear enough: you have to place weapons of the right type on each altar. The drawings on the stained glass windows in the center indicate some of the weapons to be used.

The good news is that you don't need to land specific weapons, much less magic weapons. You can simply use the rusty weapons found throughout the monastery ruins. Here is the list of weapons to use, and where to find them in the monastery:

- **Mace** : In the tomb on the edge of the cliff, southeast of the monastery
- **Battle ax** : At the feet of the divine guardian, in the sealed room. You can destroy the wall, or go through a window to enter.
- **Daggers**: On the roof, near the laser cannon (?), next to a skeleton.

You can buy basic weapons from a blacksmith if you cannot find the corresponding weapons there. Don't be afraid to use your party's magical equipment, you can collect the weapons after activating all the altars.

❖ Activate all 4 stained glass altars

The next step is to place the weapons on the correct altars. Select the weapon in the inventory, and drag it onto the altar while holding down the left click. Since the longsword is already set up, it's quite simple. If you try to place the wrong weapon, it falls to the ground and your character takes damage.

- In front of the broken stained glass window: Battle ax
- Altar opposite the broken stained glass window: Dagger
- In front of the long sword: Mace

Once the weapons are placed correctly, a wall chest will open, with a bag inside. It contains a letter and the Crest of the Dawnmaster. Keep it carefully, you will need it later.

❖ Rotate the statues of Lathander

Next, you will have to go down into the basement of the monastery, in order to reach the Githyanki nursery. Cross the area by diplomacy or force. You must access the Inquisitor's wing, passing through the captain's office, with a force field. After the confrontation with the Inquisitor (kill him if you wish) and the Queen, explore the left corridor. You will find two statues of Lathander there. To open the secret passage, you have to point them in the right direction. The statue on the left must face the end of the corridor (East), while the one on the right must be oriented towards the entrance to the corridor (West), which corresponds to the rising and setting sun respectively.

❖ Disable Light Walls

By taking the door that opened, you arrive in a long corridor with force fields and traps based on waves of light. The principle is the same each time: Destroy the energy source crystals to open the passage, then defuse the device in the middle of the corridor, or dodge the waves. The first crystal is easy to spot. The second is in a side passage. The third is below, in the chasm.

❖ Collect the blood of Lathander

After passing the 3 doors, you will finally arrive in the holy of holies, with the Blood of Lathander. But to be able to recover it, you must place the Crest of the Dawnmaster, which you normally collected earlier. If you don't have it, you're good to teleport and go all the way again.

If you try to take the mace without having the key, the big solar laser on the roof of the monastery will open fire and destroy everything, including you. Unless you manage to destroy the 4 solar generators remotely, in a limited time. They have 30 HP each. You can also evacuate the area, for example by having one character grab the weapon, while the others change areas. Letting Astarion get killed by the laser makes for a particularly fun cutscene and dialogue, even if he will disapprove.

The Blood of Lathander is a legendary morning star (+3) that is exceptionally useful during chapter 2 of the story, since you will face hordes of shadows and undead. It's a perfect weapon for Shadowheart, even if it's heresy to Shar!

Lunar Lantern

In order to continue your adventure in Baldur's Gate 3, and reach the Highlune Towers serving as a base for the cult of the Absolute, you will have to cross the shadow lands. But the places are cursed, those who venture there are killed and transformed into undead. Using torches and different light sources is enough to roam the outskirts, but you'll need extra protection before venturing into the darker places. You need a moon lantern.

❖ Where to get a Lunar Lantern?

If you passed through Malforge, and Nere was killed by you, you must have gotten your hands on a broken lantern. There are others in the same state hanging around in Chapter 1. One way to get one is to ally with Minthara against the Tieflings and Druids during Chapter 1, but that's not a good option in the majority of cases. So you have no choice, and you have to do without a lantern at the start. Use the light spell and torches until you reach the Inn of the Last Glow in the Shadowlands. It is easily recognizable by its luminous dome, and Jaheira awaits you at the entrance.

Progress the main story by speaking to Jaheira in the inn, then to the Priestess of Selune upstairs, to obtain her blessing. It will protect you from the basic effects of the curse, even without light. A big fight will follow, with an enemy attack.

After killing the attackers, you receive a new quest: Obtain a way to protect yourself from the darkest shadows. A group of minstrels awaits you on the bridge in front of the Last Glow. By talking to them, they will guide you to an ambush point. An Absolute patrol arrives, with a precious working moon lantern.

❖ Attack the Patrol and Break the Sanctuary

This step may be difficult depending on the composition of your group and the difficulty chosen. A drow drider with a lantern, some goblins and orcs arrive. Wait until they approach to order the minstrels to attack. Kill all the small enemies, and save the drider for last, he's quite sturdy with his sanctuary effect. Use area attacks to counter the effect of the Sanctuary, for example, using elemental arrows and shooting next to the Drider. Grenades, or the Cleric's Spirit Guardian also works. The objective is to break the Drider's concentration, which will allow everyone to attack him.

When he dies, the Drider will drop the precious moon lantern.

❖ Deciding what to do with the pixie in the moon lantern

By examining the lantern, a dialogue will begin with its occupant: it turns out that a pixie is trapped inside. She goes by the sweet name of Dolly Dolly Dolly. It is she who emits the light. You can decide what to do:

- Keep it locked in the lamp.

- Torture her to increase the brightness.

- Free her, hoping she will be of service to you.

The third option is obviously the benevolent choice. Dolly Dolly Dolly is a capricious fairy, even a little mischievous, but she will agree to help you. She will give you a Filigree Bell which will give the group protection against deep shadows, without having to equip anything. By using it, it will also trigger a dialogue with Dolly Dolly Dolly, to reactivate the bonus. If you ever leave the area, before returning later.

If you save the pixie, one of her colleagues will also give you a little help during chapter 3.

❖ Another moon lantern

A second way to obtain the precious lantern is to infiltrate the High Moon Towers. Upstairs, in Balthazar's room, you can find a lantern. This requires picking the door and entering a normally forbidden area. Gayle can also enhance it with a forbidden ritual.

Help Dammon

It's a real moment of happiness to meet your first blacksmith in Baldur's Gate 3. It's the opportunity to sell your equipment and, above all, to buy new magical weapons and armor. Dammon has plenty of good things for you, and he's at the heart of an important quest for one of your companions.

❖ Dammon's position during Chapter 1

This merchant is located in the Druid Grove, in the section populated by Tieflings. Her shop is opposite that of Auntie Ethel, this brave woman. Don't feel obligated to loot the magical weapons and armor on display at his stall, one way or the other, you can pick them up later without risking being spotted.

He is the one we must go see in the company of Karlach, in order to give him some Infernal Iron, to repair his heart.

Be careful, if you decide to ally yourself with Minthara and the goblins, Dammon will die. This will compromise this companion's quest line, which may leave you permanently abandoned in any case.

❖ Where to find Dammon during chapter 2?

If you have eliminated the goblins and saved the Tieflings, Dammon will move with his peers. It can no longer be found in Druid Grove.

Instead, he can be found in the Shadow Woods, in the Highmoon Towers region. Join the Utimeglow Inn, which is easy to spot with its huge glowing dome.

Dammon is to the right of the inn, in the stables, with the oxen. He offers new products, and above all, he can repair the heart of Karlach once again, with a second piece of Infernal Iron.

❖ *Where to find Infernal Iron?*

The developers have been generous, it is possible to find Infernal Iron almost everywhere in Baldur's Gate 3. Even if you miss a piece, you can always find another. If you sold one to a merchant, you can normally buy it back, steal it, or loot it from its corpse.

◦ *Chapter 1*

● **Village in ruins**: In the locked forge, the first building on the right when entering the village. Goblins are hiding on the roof. You can force the door, or go through the hole in the ceiling after destroying the cobwebs.

● **Treasure room - Goblin Camp** : In the huge pile of chests full of gold and treasure, behind the locked gate, at the very back of the area. He is at the top of the stairs, behind the warlord and his throne.

● **Zentarim Hideout (X:282. Y-167)**: The entrance is in a cabin west of Waukine's Rest, the burning inn. You must kill the guard at the entrance or give him the password. The piece of iron is in a locked chest in the area.

● **Malforge (X:661 Y:368)** : Towards the entrance to the area, you will come across two Duergars, a merchant, and another, named Kith, who asks you what you see while inspecting the rubble. You must succeed in 3 skill rolls, History (d10), Investigation (d10 and Perception (d10), remember to activate Assistance and save before. If you succeed in all three rolls and report to him what you have noticed, he will give you the Infernal Iron.

Thisobald Thorm

Madness has gripped the Shadow Cursed Lands, with the undead strangely fascinated by the passions they had while alive, such as alcohol or gold. Thisobald Thorm is a brewer and bartender, not necessarily as bad as he seems, when you know how to handle discussion, or combat. This is certainly one of the most interesting bosses in Baldur's Gate 3.

❖ Dialogue and peaceful solution

Be careful, when entering the brewery, Thisobald Thorm will automatically start a dialogue with the nearest character, and you will not be able to change it. Make sure this is the character you want.

If you refuse to drink with Thisobald, the fight will immediately begin. It would be a shame to miss this dialogue.

- We advise you to use Karlach, who as a Barbarian is perfectly capable of handling the situation. That is, drinking immense quantities of filthy beer with special dialogue options. She is entitled to Constitution rolls with advantage. With any luck, she can get the boss drunk while extracting information from him.

- The alternative is to use a character with a good Deception score to pretend to drink. A Rogue or a Bard will do the job perfectly.

You can ask Thisobald questions after drinking at least once, then you are free to focus on the drink, or on the questions. You can also mix the two to make Thisobald more malleable.

- If you ask questions about Kéthéric Thorm and Chantenuit, about the nature of the relic and its prison, Thisobald will give you information that he was not supposed to reveal. If he refuses to answer, have a drink, and start again. By focusing on this subject, the curse will activate, and Thisbald will die by himself, without having to fight.

- If you insist on drinking, the Constitution (or Deception) saving throws will get higher and higher, but there are several advantages. Not only will the boss respond more to you, but above all, if you interrupt the conversation and the fight begins, he will be much easier to kill. Alcohol is not his strength, but his weakness. You can push him to the point of unconsciousness, which makes him extremely vulnerable.

❖ Fight against Thisobald Thorm

- Thisobald has relatively high armor, 16 AC, almost 300 health points and above all: he is completely immune to physical and thunder damage. It is very difficult to damage it in these conditions.

- The obvious solution is to use elemental, necrotic, and radiant damage. The Cleric's Crusader Aura, Damage Curse, Occultist's Curse, weapons and equipment that add elemental damage are all ways to damage him even while attacking normally. Elemental arrows and bombs are also formidable. And of course, elemental spells are going to hurt him a lot too.

- The small problem is that the boss will unleash an elemental attack around him, or a breath in front of him during his turn. The element used by the boss depends on the last damage received, so the result may vary. If you have to choose an element, it seems like Fire is the best option, as it's the one you're most likely to resist. Radiant and force damage don't seem to buff the boss.

- If you lack elemental damage or survival, you can approach the fight differently, and delay. Indeed, each time the boss plays his turn, he will drink alcohol, which will increase his blood alcohol level, and therefore his penalties. After a few turns, he will collapse, which will impose -5 AC on him, and will cause his immunities to physical damage to disappear.

- In this case, space out your characters, and simply keep the tank.

- Remember to eliminate the zombies that will also attack you during the fight. They're not very dangerous, but they have quite a bit of health. You can assassinate some of them discreetly before the fight.

- If you're having difficulty with this fight, taking a detour through the Mountain Pass to collect the Blood of Lathander will help you immensely. The mass will blind him during the fight, its damage is immense, and the ray of light can easily kill his zombies.

❖ Awards

In addition to a lot of experience, and a lot of alcohol, you will get a key. A mass of rat killers, and the journal requested by the soul-torturer elf is hidden under a board at the back of the bar.

Thisobald Thorm also has a laboratory, at the back of the brewery, with more loot and information.

Parents d'Arabella

Baldur's Gate 3 is not for the faint of heart, and you're usually just one response away from seeing a grotesque or tragic scene unfold before your eyes, and children are not spared.

❖ Where to find Arabella during chapter 2?

If you saved Arabella's life during the confrontation with the serpent of Kagha, during the first chapter, then you saved the Druid Grove, then she will return during the second chapter.

You meet him while exploring the cursed shadow lands, between the cemetery and the masons' guild, to the west of the map.

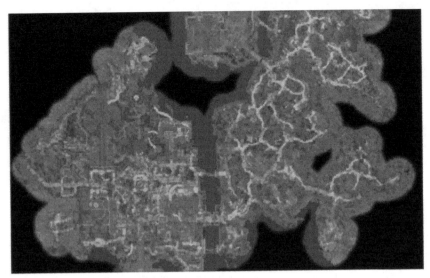

It is strongly recommended that you tell Arabella to go to your camp, and accompany her there. There's a good chance she'll run into trouble and die if you send her to the Last Glow Inn.

Once at the camp, talk to Arabella and Blight, who will become friends which is rather unexpected. Different dialogues will appear in the camp over the days, if you like this type of interaction.

You can rest, or go looking for Arabelle's parents afterwards.

❖ Position of Arabella's parents

Not a big surprise in the Cursed Lands, but Arabella's parents are dead. As far as we know, it is not possible to save them. You can find their corpses on a hospital bed, in one of the wings of the Healing House, in the west of the city. The Halsin quest also asks you to go there.

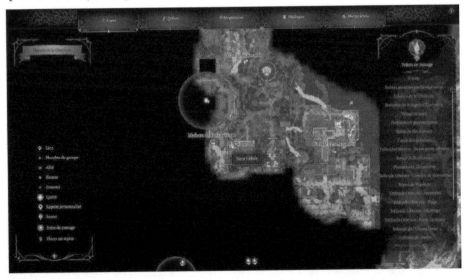

- A corrupt nurse is nearby. You can convince her that they no longer need care, in order to buy goods from her. You can also kill her to avenge them.

- You can use the Communicate with the Dead spell on Arabella's father to obtain some information on the circumstances of their death, and on Zevlor's disappearance.

- The mad surgeon is also waiting for you with his nurses, a little further in the area, even if killing him doesn't change anything regarding Arabella's parents.

- By returning to talk to Arabella at camp, you can decide to hide the news from her, or to tell her the truth. If you want a more positive outcome, it's best to be honest with her, and encourage her to accept this tragedy.

- The next day, Arabella should tell you the fate of Lianes, in order to thank you. As well as a ring capable of casting an Illusory Dagger spell.

- She should also mention that she intends to leave, in order to cultivate her gift for magic. Arabella will then stay at the camp, until Chapter 3. She will then end up disappearing, leaving you a thank you note.

Gauntlet de Shar

If you like dungeons full of monsters and twisted trials, you're going to have fun in the Temple of Shar in Baldur's Gate 3. Here's a list of all the most important items.

❖ Chapter 1: Ancient Temple of Shar

If you are still in the first chapter of the game, and after freeing Nere, while visiting Malforge, you see the Temple of Shar in the distance, it is normal to try to enter it. But it's impossible since this map and chapter. You will definitely have to visit the places during chapter 2.

❖ How to enter the Temple of Shar?

If you're looking for exactly where to go to enter, here's its location on the main map of Chapter 2. You'll need a Moon Lantern, or similar tool, so you don't die from the shadows along the way.

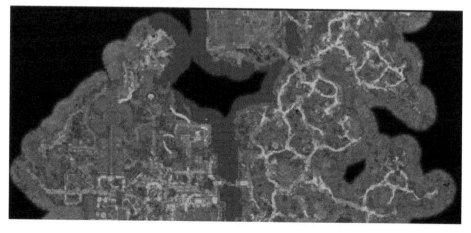

❖ Raphael

As you approach the entrance to the Mausoleum, that good little devil Raphael will make his return, to tell you about a powerful enemy locked inside: Yurgir. It's best to have Astarion in the party, since he has some questions for him as part of his personal quest.

The answers you give don't matter, since one way or the other, it's almost obligatory to kill Yurgir.

❖ Mausoleum

You start with the Great Mausoleum, where Ketheric Thorm was buried. It is advisable to read the books and documents present, they provide clues about what happens next, and about the motivations of your enemy.

To progress in the mausoleum, you will have to cross a room full of formidable traps. Fortunately, there is an easy way to defuse them: click the buttons under the murals in the correct order. They transcribe the journey of General Ketheric:

1. Highlune Towers

2. Grief

3. General (at the other end of the room)

❖ Edge of Shadows

After taking the flying platform, here is the first test truly linked to Shar. The principle is very simple when you understand the principle:

Venture into the 4 rooms, in the 4 corners of the room (watch out for the traps), then activate the lever they contain each time. Sending a lone character, who jumps over the pressure plate at the entrance, and avoids the circular trap in the center, should prevent you from taking too much damage.

Each lever will bring down two lamps, which you can turn off by clicking on them. Technically, it is possible to turn them off in other ways, but this requires using consumables and spells.

Then, send a character detached from the rest of the group to approach the statue of Shar in the center of the room. Avoid the blue lines on the floor, imagine they are walls. Then activate the console.

❖ Shar's Temple of Trials

You then arrive in the large hall of the temple. A battle will be fought between undead and shadows. Your priority is to destroy the shadow portals, in order to cut off the flow of reinforcements.

To complete the trials, you will need to obtain a total of 4 Sooty Gems scattered throughout the area. By going to the West, you will be able to start the normal tests Going east, you will come across Yurgir, the big boss linked to Raphaël. It holds one of the famous gems that you will need.

It is strongly recommended to have Shadowheart in your party in this area, since it is closely linked to his personal quests.

❖ Balthazar

By following the corridor to the West, you will come across a rat giving a small quest, if you use communication with animals, and you kill it twice.

By continuing your path, you will trigger a big battle between the undead and the shadows. After destroying numerous portals, you will meet Balthazar, an awakened soul. He will send you to recover Chantenuit, the relic of General Ketheric (that was a long time ago). He will also give you a summon bell from his brother, a big, very strong zombie. It can be used once, but only in the Temple of Shar. If you intend to confront Yurgir directly, his help will be welcome.

You can temporarily follow Balthazar's orders, and even trade with him. You can also fight him, but be prepared for a big fight. He summons many undead, and he uses the fearsome Death Cloud spell. Killing him now gives you the advantage of surprise and terrain. But don't worry, you will have the opportunity to kill him later at the end of the temple. Or to ally yourself with him.

Do not be afraid to shed your blood, or to have Shadowheart shed his. Ideally, discretion should be exercised. Avoid the patrols, collect the test key in the small room, then jump out the window to avoid an enemy and the grate. Open the final gate with the key, then collect the gem.

❖ Leap of Faith Test

You just have to walk, or jump on the path camouflaged by shadows. If you have trouble seeing it, change the game brightness in the options to increase it to maximum.

❖ Doubling test

A great classic, always fun. The game will create a copy of your party, with the same equipment, spells and abilities. The enemy group will then ambush you on the stairs. A good way to make the fight easier and to completely undress your group before starting the ordeal (the shedding of blood), you should be familiar with romances. Approach the area quietly, then re-equip all your items. Elden Ring players are used to it. Discreetly go around the area to the right, in order to try to take the enemy group from behind, and by surprise (a perception roll is required). Be careful not to get pushed into the void!

❖ Silent library

This is the only test that is a little difficult, since it requires you to use your brain a little. Enemies will throw silence bubbles at you, but moving to get out of them is enough. The objective of the area is to collect the books from the library shelves, then place the correct one on the altar in the center of the next room. Be careful, these shelves are all trapped. If you make a mistake, this will trigger a ground trap. It is possible to open the door using brute force: activate the altar in a loop, with each book.

The good book is on the second shelf, to the north of the room, the Teaching of Shar: the night song.

This will open the large door, and give you access to several chests, with equipment Take care to keep the Spear of Night, you will need it. There is no gem in this case.

❖ Yurgir

As you venture into the eastern section of the Temple of Shar, you will see an eclipsing beast resembling a black panther, which will flee. She seeks to lure you into the trap of a formidable devil: Yurgir. When entering the room where he is, you will be in a very bad position, with the beast at your back, a series of fairly strong legionnaires on the platform above, and above all, Yurgir with his heavy crossbow and his powers. He will throw bombs at you from his ledge. And the most painful thing is that he is capable of making himself invisible.

Here are some ways to work around the problem:

● Having surgery by Volo.

● Potions/scrolls/detect invisibility spells.

● All area techniques, such as Fireballs, Burning Hands, Guardian Spirit and many others, can force him to reappear.

You can also pass discreetly through the southeast corner of the area, in order to start the fight at height. This will make targets more accessible and easier to hit. You can also push them down.

An alternative with excellent dialogue skills is to analyze the contract/curse that is holding Yurgir back. You can then push him to kill his beast, his servants, and ultimately, himself. Driving a boss to suicide is the ultimate class for a Bard!

Remember to pick up your equipment and especially the last shadow gem.

❖ Ancient Altar and Chantenuit

Once in possession of the 4 sooty gems, return to the main hall of the temple to activate the altar with 2 gems. This will activate the platform, which will take you to the second section of the area.

You will encounter a second altar, asking for the last 2 gems this time. This will give you access to a fast travel point, and to the final area of   the temple. But be careful, don't enter yet. This will trigger the rest of the scenario, which will make a lot of quests and characters unavailable. It is better to search the entire region, including the Hautelune Towers first. Entering the pool should be the last thing you want to do in this chapter before facing Ketheric Thorm.

Githyanki Nativity Scene

One of the most choice-heavy passages in Baldur's Gate 3 is the Githyanki Crèche, especially if you want to keep Lae'Zel in your party. You have to find the right balance to obtain her approval, and rally her to your cause, without ending up dead or brainless.

❖ How to find the Githyanki Nativity Scene?

You have to enter the Mountain Pass region, which triggers the start of Chapter 2. It is better to wait until you have the problems in the Grove of Druids and Goblins, one way or the other. The two passages to the west of the map are viable, but it is better to take the passage to the northwest, in order to come across the patrol and the dragon, in the company of Lae'Zel.

● Once in the Mountain Pass, go to the Lathander Monastery in the North. You can take the cable car, or go down along the cliff.

● When you arrive at the monastery, the front door closes. You can simply go through the ruins on the right, by climbing then jumping below for example. Next, go down the stairs behind the statue.

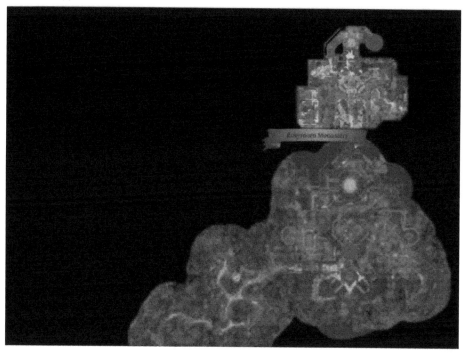

❖ Infiltration of the nursery

You can let Lae'Zel negotiate with the guards at the entrance to the nursery. It is also possible to take matters into your own hands. Using the Disguise spell to take on the appearance of a Githyanki tends to open up more options during dialogue.

It is advisable to visit the different secondary sections of the nursery, before entering the captain's office and going to see the inquisitor. This may no longer be possible afterward. Go see the seller to the east of the area to get your hands on some items exclusive to the Githyanki.

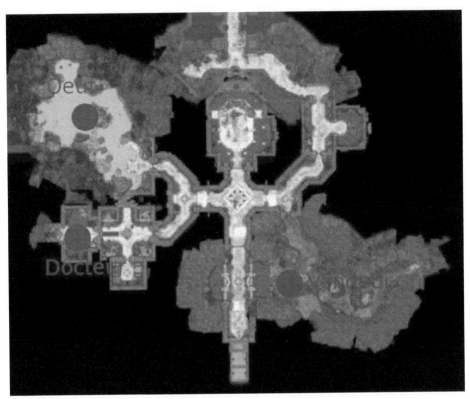

You can also have fun defacing the queen's painting. Strangely, Lae'Zel doesn't disapprove. In addition to being quite funny, it will distract one of the guards.

❖ Infirmary: Remove the larva with the Zaith'Isk

Upon entering the infirmary, you will be able to speak to an awkward Githyanki scientist who is in charge of extracting larvae with her machine. As suspected from the beginning, the procedure is fatal. It is nevertheless interesting to play the game.

- You have two notable choices: Let Lae'Zel go first, which she approves of. This allows you to observe the procedure and intervene. If you have an excellent persuasion score, you can convince her that the procedure will kill her and to stop everything. This is later useful in convincing her to abandon the queen.

- A much more interesting, but dangerous, alternative is to climb on the device yourself. Remember to save the game beforehand. You must succeed in 3 saving throws to resist the Zaith'Isk, this will grant you the "Awakened" bonus allowing you to use your Illithid powers with a bonus action. This makes them much easier to use. In return, failing the rolls will earn you a permanent penalty to Intelligence and Constitution.

After the procedure fails, you can try to bullshit the friendly doctor. But it's better to kill her, with the Githyanki present in the infirmary. Remember to collect all the larvae in jars stored in the area.

❖ Incubator: Obtain a Githyanki Egg

- In the mountain pass, you must have come across a scholar offering to find her a Githyanki egg in the crèche. At first glance, Lae'Zel is hostile to this idea, but by carrying out a few dialogues, the situation can be managed later.

- Once in the incubator area, you will arrive in front of a pool of acid full of traps It is better to separate a member of the group, and use mobility abilities to reach the person responsible for the range, at the top of the central rock. Using large jumps from rock to rock, Misty Step or Flight are all viable solutions.

- Tell the leader that you are going to find a nursery that is better suited for this egg, in order to convince him to hand it over to you peacefully, while gaining approval from the rest of the group. He'll even give you a pair of acid-immune boots. It will allow you to reach the egg without taking damage.

- Bringing the egg back to the scholar allows you to obtain gold, and this gives access to a sequel, with some negative but interesting repercussions during chapter 3.

❖ Captain's Office

Things get tougher when you decide to progress through the story and zone. A force field will block your path. It is almost obligatory to speak to the captain, to pass peacefully.

- You can tell him there is a traitor, but without revealing his identity. Then insist that you can only tell the Inquisitor about it.

- It is also possible to reveal the relic directly to the captain. She will try to seize it, but without success. She will then send you to see the Inquisitor too.

❖ Should we show the relic to the Inquisitor?

The Inquisitor already knows your identity, and he already suspects that the relic is in your possession. It is impossible to fool him. You can choose to show the relic to the inquisitor, or engage in combat immediately. This second option is more interesting in our eyes, since you will inevitably have to face it later. It also has some great items. Killing him beforehand reduces the chances of things going bad with Vlaakith and Lae'Zel turning against you.

❖ Dialogue with Vlaakith

Whether you killed the Inquisitor or not, the dialogue will automatically begin with a projection of Vlaakith, the Lich Queen.

- If you want to improve your relationship with Lae'Zel, avoid being disrespectful. It is best to bow down and agree to consider what Vlaakith asks of you, without openly rejecting it. Don't behave too slavishly either, to avoid disapproval from your other companions.

- Don't doubt his ability to kill anyone though, that would lead to an instant Game Over. It's still pretty fun to watch though.

❖ Relic: Should its occupant, the Night Visitor, be killed?

Unless you are killed by his wish, Vlaakith will open a portal to the interior of the prime. You can then enter the cave alone, in order to speak to your visitor, or nocturnal visitor. Take advantage of this to extract as much information as possible especially about Vlaakith lying to her people, and manipulating them.

When he puts his life in your hands, refuse to kill him. You may choose to do this out of curiosity, but it won't lead to anything good anyway.

When you leave the cave, Lae'Zel will question you, and you must tell him the truth The best option then is to convince her that Queen Vlaakith is manipulating and deceiving her people. That there is no ascension. If she used the Zaith'Isk, she is easier to convince by choosing the dialogue option mentioning the fact that it was going to kill her. Otherwise, you will probably have to make a persuasion roll.

❖ Leaving the nursery

By taking the prism exit portal, you will return to the Inquisitor's room. If he is still alive, the fight will inevitably begin. All the other occupants of the nursery also became hostile. You can take the time to slaughter all the other Githyanki, in order to stock up on experience and valuable items.

Also take the time to rotate the statues in the corridor to the West, in order to obtain access to the Blood of Lathander.

❖ Kith'rak Voss

During your next long rest at camp, a group of Githyanki will come to visit you overnight. The Kith'rak Voss will reveal himself to be indeed a traitor, who fights against Vlaakith. The best option is to push Lae'Zel to spare him, and join their cause. It is by far the most profitable and interesting later, with a legendary weapon at stake. But if you want to play a villain, with catastrophic choices, you can encourage Lae'Zel to stay loyal to his queen, and eliminate traitors.

The Forgotten

The serious things begin with the second chapter of Baldur's Gate 3, the identity of the Absolute and the plans of Kéthéric Thorm end up being revealed. But surprises also await you, with unexpected zone.

❖ How to enter the oblivion?

There is no shortage of options for reaching this area, intentionally or not. The majority of them are in the prison beneath the Towers of Hautelune. You can jump into the large corpse pit to the south of the area, at the risk of taking fall damage. The alternative is to destroy the walls at the back of certain cells, in order to reach a secret corridor. To the west, you will reach a cliff with its holds, allowing you to descend into the oubliette without breaking your knees.

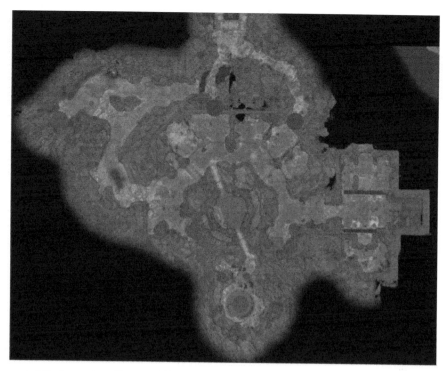

❖ Noises in the walls

An alternative method is to climb the beams high up on the ground floor of the Hautelune Towers. As you navigate near the ceiling, you will reach a cracked wall. You can click on it to interact with it. This will trigger a dialogue, with the possibility of removing your arm or not. By persisting, you will eventually speak with the Absolute itself. As you continue to keep your arm inside, the tentacles will forcefully pull you into oblivion. This event is linked to the quest for noise in the walls of the towers.

❖ What to do in the dungeon?

At the moment, there is nothing interesting to discover in the area. You are in a subsection of a larger area, which you do not yet have access to. It is impossible to reach it from the oubliée. It will be necessary to make the Gauntlet of Shar, and decide the fate of Chantenuit, then face Kéthéric Thorm. In the second part of the area, you will find the she-devil that Mizora asked you to free, Tiefling prisoners, as well as clues about the Grand Duke. Not forgetting to mention, that the end of the chapter will take place there.

This can be considered a preview of what awaits you next, it won't come out of nowhere. You can loot the area to collect a few items, and gain some experience, before looking for the exit.

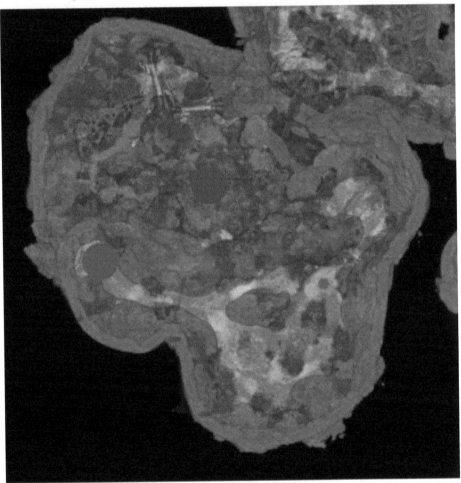

❖ How to get out of oblivion?

This is one of the rare areas in the game blocking fast travel and returning to camp. If your group is in bad shape, and you have landed in the area unintentionally, do not hesitate to use the consumables.

Fortunately, the area isn't very big, nor are there too many enemies. The only major obstacle is a group of leaping hooked horrors. Take the passage to the east, which then turns southwest, to reach the foot of the cliff allowing you to exit the area. It leads into the secret corridor of the prison.

Elminster

If Elminster Aumar were free to intervene in Baldur's Gate 3's story, he could probably solve the majority of our problems with one hand behind his back. It must be said that the creators of the license made him a character of absurd power, and he is probably the most powerful mortal in the universe. Fortunately, he is more there to guide the story in the direction desired by his goddess, even if it is not necessarily the right one, as we will see. For information, he was also present in Baldur's Gate 1 and Baldur's Gate 2 Throne of Bhaal, with a similar role.

❖ *Where to meet Elminster?*

The famous wizard can only be encountered at the end of the first chapter, when you enter the Mountain Pass, or when you pass from Malforge to the Cursed Shadowlands. In summary, it is definitely on your way. He appears in the form of a weary traveler.

But to talk to him, you must have recruited Gayle, and have him in your group, or your camp. If you gave free rein to your dark impulses, or if you refused to give him magic items to cure his "little health problem", and he left the group, you will not have any possible interaction with Elminster.

❖ *Elminster and the Wizard of Waterdeep*

After a brief dialogue, Elminster will invite himself to your camp to steal your cheese and incidentally, talk to Gayle about the stupidity of his past actions, as well as their present repercussions. It will calm Netheril's Orb of Destruction, and put it under your control: you no longer need to sacrifice magic items regularly, and a button appears on the list of shortcuts. He asks Gayle to sacrifice her life, causing the orb to explode at just the right moment, in order to destroy the Absolute, while eliminating the threat it poses to the fabric of magic. It's not very nice of him, but Gayle deserves it. You can also comment on it with humor, even if Gayle will disapprove.

If you press this button in combat, a cutscene will trigger, and Gayle will explode, which will automatically trigger a Game Over, even if he is not in the same area, or the same region as the rest of the game. band. This will also give you the opportunity to trigger two additional endings to the game, one at the end of Chapter 2, and one at the very end of the game, if Gayle is in your party. After explaining all this, Elminster will bow out.

❖ *Chapter 3: Elminster Reappearance*

If you agree to accompany Gayle to the magic chests under the Magic & Sorceries shop in Baldur's Gate, in order to read the Annals of Karsus, he will discover key information about the crown.

During a Long Rest in the following days, Elminster will visit you once again at camp, to ask Gayle to go to the temple in Lower Town. An interview with the goddess Mystra herself awaits him. You should take his advice and go there.

❖ Is it possible to kill Elminster?

The answer is complicated. Indeed, Elminster only has 96 life points, and nothing prevents you from attacking and killing him if you feel like it, even if this can have serious repercussions on Gayle's story, in depending on when you do it. But of course, it takes more than a simple ambush to kill the most powerful magician in the kingdoms. This is not the real Elminster you encounter, but a simple magical simulacrum, created with snow and controlled remotely. When you kill it, it melts and all that's left is a puddle. There's no notable loot either.

In summary, apart from the pleasure of killing this horrible bearded "Mary-Sue" created by the author, it brings nothing, and we cannot really consider having defeated him.

Guardian of the faith

Chapter 2 of Baldur's Gate 3 offers radically different areas, with on one side the Cursed Shadowlands, and on the other, the mountain pass with its monastery full of light, even too full of light. Different obstacles await you there, including the Guardian of the Faith, if you wish to obtain the Blood of Lathander.

❖ How do I deactivate Faith Keeper?

As you climb upstairs in the monastery, you will discover a door covered in a golden force field called Guardian of the Faith. Don't bother trying to disable it, it's impossible as far as we know. But there is no need for it anyway, you can get around the door in at least three ways. It's a bit like putting an armored door on a straw house.

The simplest and most reliable method, which avoids having to micromanage the group's movements, is to destroy the exterior wall of the room. Retrace your steps, and inspect the right wall of the room. It has a life bar. Destroy it with a blunt weapon for example.

You can also jump through the window in the room with the altars, in order to reach the ledge outside. By following it, you will reach the window in the room behind the indestructible door. You can also jump inside from the roof of the monastery, there is a hole.

❖ How to pass the Divine Guardian?

Once in the room, you will see a summoned Guardian of Faith, protecting a ceremonial ax that you need. You can attack it, or simply send a summoned creature into range. All damage dealt by the Guardian of Faith is deducted from its hit points. Suffice to say that he will quickly commit suicide. If you don't have a summon, send a sturdy character like Lae'Zel or Karlach to absorb the damage.

Once the guard is dead, you will discover that the force field of the door is... still active. All you have to do is get back out, which is easy if you've demolished a wall. All that's left to do now is solve the puzzle at the altars, before heading to the Githyanki nursery.

Balthazar's Coin

Baldur's Gate 3 isn't going to hold your hand by telling you every time where to go and what to do, but it does tend to offer several ways to get there. As Minthara invites you (if applicable), following in Balthazar's footsteps is fundamental during chapter 2, in order to reach its conclusion, and to overcome the immortality of Kéthéric Thorm.

❖ How to enter Balthazar's room?

We can consider that this problem is divided into two phases, the first consists of entering the room, the second, in the secret room. Let's start with the bedroom. There are at least two ways to enter:

Passing through the exterior of the building, you can reach the balcony shared by Balthazar and Kéthéric's rooms. You can climb up via the vines and ladders outside the building. Spells like Fly and No Mist are also a big help. You must then unlock the door, which has a D30, which is the maximum in play. The easiest way is to use a level 2 Unblocking spell, you may have it with Gayle, or on a parchment. Otherwise, you will have to make repeated attempts with Astarion or another character, and have a little luck. The Assist spell, or Cat Agility, will help you greatly.

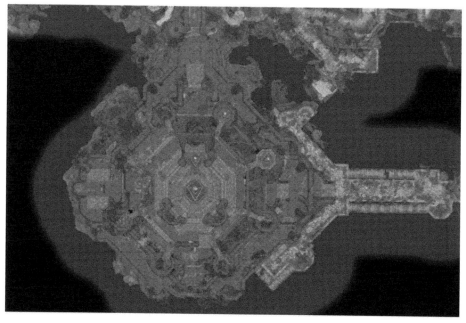

The alternative is to go through the interior of the building, climbing to the first floor, then infiltrating the restricted area. You can also convince Z'rell his apprentice that you are real admirers of the Absolute in order to obtain the key. However, it is quite difficult, since a guard patrols the area, as well as a flying eye serving as a sentry. Try to kill the flying eye out of sight, it can be killed instantly without sounding the alarm. It's more difficult for the guard. A Major Invisibility spell is the easiest way to get through without being spotted and open the door.

❖ How to open the secret room in the necromancy laboratory?

Once in the room, remember to search everything and pick up useful objects, such as the Lunar Lantern. If Gayle is present, he can use it with the ritual circle to create an improved version, at the risk of angering Mystra. It's better to let him take care of it and destroy the ritual in order to obtain a permanent Concentration bonus.

You can also find different documents offering information on Balthazar and Kéthéric. But what really interests us is solving the puzzle in order to open the passage leading to the secret room. Here are the steps to follow:

1. Click only on the books at the top right of the shelf.

2. Collect the heart placed not far from the altar.

3. Place the heart on the ancient altar by interacting with it.

The library will open, giving you access to the secret room.

That being said, you don't really need to venture into Balthazar's room to tackle the rest. You must go to the large mausoleum in the north of the region, in order to reach the Gauntlet of Shar. You will be able to meet Balthazar in rotten flesh and bones, in order to assist him or not, in his search for Chantenuit.

Balthazar

Life would be bleak in the Forgotten Realms without Necromancers like Balthazar to summon the undead and concoct evil rituals. You will have several opportunities to face Balthazar in Baldur's Gate 3, and even to temporarily ally yourself with him.

❖ Where is Balthazar located?

If you started by searching the Towers of Highlune, you can enter Balthazar's room and laboratory to loot his goods and research. The tower guardian will also ask you to go see him to help him recover the relic of Kéthéric Thorm.

To find this brave man, you must go to the Mausoleum to the north of the city, then progress in the Gauntlet of Shar, in order to reach the room in which he is hiding with his undead

❖ Gauntlet de Shar

After passing the first test in the temple, you will reach an Ancient Altar with a battle against shadows. Go west, you will trigger a second battle between the undead and the shadows. After destroying numerous shadow portals, you will meet Balthazar, an awakened soul. He will send you to recover Chantenuit, the relic of General Kéthéric. He will also give you a summon bell from his brother, a big, very strong zombie. It can be used once, but only in the Temple of Shar. If you intend to confront Yurgir directly, his help will be welcome. You can temporarily follow Balthazar's orders, and even trade with him.

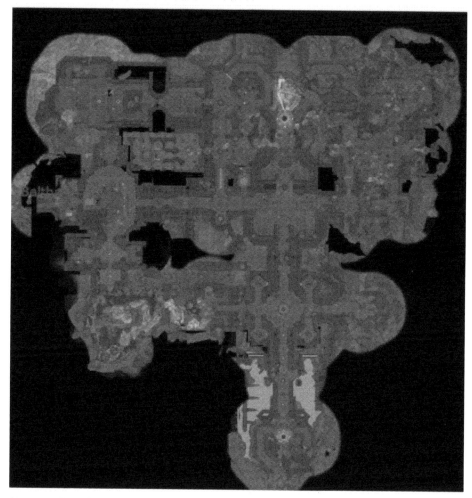

You can also fight him, but be prepared for a big fight. He receives help from the undead in the area, and he uses the dreaded Death Cloud spell. Killing him now gives you the advantage of surprise and terrain. But don't worry, you will also have the opportunity to kill him later, at the end of the temple. Or to ally yourself with him.

❖ Ombregrise

If you activate the two Ancient Altars, and you manage to enter Shar's domain, Balthazar will appear. He took advantage of your efforts to get through, the deceiver! He then flies away, leaving you to fend for yourself. You can find him at the end of the area, alongside Chantenuit, who is a prisoner of the ritual. You now have two options:

- Confront Balthazar and kill him: This is the strongly recommended option if you have Shadowheart in the group, since she must decide the fate of Nightsong.

- Maintain your cooperation with Balthazar, and let him kidnap Chantenuit to bring him to Kéthéric Thorm. It's a rather original choice, even if the result will not necessarily be what you hope for. We will detail this a little further.

❖ Fight against Balthazar

Many players face Balthazar in Shadowslope, when Nightsong is present. It is also during this occasion that Balthazar is the most powerful. Indeed, he will instantly summon a veritable horde of undead, before teleporting to the top of the cliffs with Misty Step. Then, he will inflict the Death Cloud/Withering and Misty Step combo on you, in order to change position.

Here are some tips to make the fight much easier:

- Positioning invisible or hidden characters on top of cliffs before engaging in combat yourself, or speaking to Balthazar will help you greatly.

- Having a character with Counterspells (level 3 spells) in reserve will prevent Balthazar from using his most dangerous spells. Block Undead Summon or its Death Clouds as a priority.

- A popular and hilarious technique is to push Balthazar into the void (like Auntie Ethel), with an action, spell or consumable. In return, you will lose his loot.

❖ Booty

As a good Necromancer, Balthazar will leave you a Circle of bones offering your undead servants within 6 meters resistance to physical damage. It also allows you to cast the Level 3 Animate the Dead spell once per day.

❖ Let Balthazar remove Chantenuit

If you decide to be evil, or ally yourself with the Absolute, you can let Balthazar take over Nightsong. The rest of the events then take a slightly different turn: beware of spoilers.

You meet Kéthéric Thorm at the top of the Hautelune Towers. And he is grateful for your help. Unfortunately, the situation quickly takes an unfavorable turn, as he performs a ritual aimed at locating the relic in your possession. He then calls on the Absolute and its giant tentacles to control you. This gives an automatic defeat.

You regain consciousness in an organic cell, in the middle of the Illithid hive, the other section of the Oubliettes. Your companions are also prisoners. Balthazar is about to experiment to get the relic from you. You must destroy your cell, free your teammates, then confront Balthazar, or risk dying yourself. In summary, a long-term alliance is impossible.

The rest of the events are similar to the other routes in the game. With a confrontation against Kéthéric Thorm asking you to free Chantenuit from the ritual The funny thing is that you can always ally with Chantenuit and Jaheira afterwards claiming that this whole thing was your infiltration plan, and it worked perfectly. You can always take advantage of this to cooperate with Loarrakan during chapter 3, to betray and deliver Chantenuit a second time.

Ketheric Thorm

After having massacred a band of lousy goblins, or a band of quarrelsome druids, it is easy to underestimate the enemy during the second chapter of Baldur's Gate 3. But this time, your antagonist publicly exposes his "plot armor" which he We will have to cope for a good part of the next few hours.

❖ Meeting with Ketheric Thorm in the throne room

By going to the Towers of High Moon, you will be able to follow an audience granted by Ketheric to a band of goblins. This is the opportunity to confirm his invincibility. Technically, the game allows you to draw your weapons and attack, but it would be incredibly stupid to seriously expect to win this fight. Along with tons of high level enemies present, Ketheric is very literally invincible. You cannot inflict damage on him. It is better to wait, and tackle the problem piece by piece. Talk to the Guardian to get information from him, and search the place, there are plenty of very profitable quests to do.

Ketheric will go to the roof of the Highlune Towers, you can make your way to him by killing his servants, but once again, this will be of no use for the moment. You can't even try to kill him by pushing him into the void, he is immune to movement effects. Come back when you have discovered its secret.

❖ How to end Ketheric Thorm's immortality?

The region is full of documents and clues about Kéthéric Thorm, not to mention several objects specifically designed to combat the undead and shadows. The Blood of Lathander, present in the Mountain Pass Monastery, is absolutely fantastic in this area. But none of this will help without making Ketheric mortal again.

The solution is to go to the Great Mausoleum in the northwest of the Cursed Lands region. As you progress through the area, you will reach the famous Gauntlet of Shar, and its many trials. You can discover them in our dedicated guide below. The important point to remember is not to cross the last passage, before having completed everything you wanted to do in this second chapter. A warning message will notify you when you are at the point of no return.

❖ Deciding what to do with Chantenuit

Passing the point of no return, you will enter Grayshadow, the domain of Shar. It is strongly recommended to have Shadowheart in your group, otherwise she risks sulking during the rest of the adventure. This will also give you an additional option. At the end of the area, you will find Balthazar, at least if you have not killed him before, and Chantenuit, an Aasimar.

Here are the three possible choices and their consequences:

● Kill Balthazar and save Chantenuit. If she is present, you must convince Shadowheart to spare her. This is by far the best option. It makes Ketheric vulnerable, and you will have the assistance of Jaheira and the Harpers to attack the tower. At least, if you managed to defend the inn.

● Let Shadowheart kill Nightsong with the Nightspear. This has significant repercussions on Shadowheart, but also on Jaheira, who is killed during the assault The advantage of this method is that Ketheric's immortality disappears for good. If you want to play a villain, it's here.

● You can let Balthazar kidnap Chantenuit to bring her to Kéthéric in your company. The rest of the events will change a little, but you will still have a way to progress, and to kill Kéthéric afterwards. It's counterintuitive, but this method allows Jaheira to survive, and even ally with Nightsong later.

❖ Can we ally with Ketheric Thorm?

The question is legitimate, and many players thought it was a real possibility. The answer is unfortunately: no. By delivering Chantenuit, you find Kéthéric at the top of his tower. He then undertakes a ritual to detect the relic in your possession, which is quite unfortunate. You are then taken prisoner. This allows you to skip directly to the first fight against Kéthéric Thorm, but instead, you will have to escape from the Illithid prison.

You can also try to persuade him at the top of the Hautelune Towers, by telling him that his wife would not have approved of his actions. You cannot convince him to stop the fight, but this will allow you to continue your attempts during the second encounter, which will push him to suicide. This is not always desirable, however, as it will complicate the fight.

❖ Battle at the top of the Hautelune Towers

As you approach Kéthéric, a dialogue will begin. You can try to persuade him to redeem himself, a bit like Darth Vader, but that won't accomplish anything yet. If you have freed or killed Chantenuit, the rest of the events await you at the top of the Hautelune Towers. If applicable, Chantenuit will be present. There is no need to worry about its survival. If she drops to 0 life points, she recovers everything the next turn. You can let it tank.

The combat is rather simple, kill the enemies in the area, then concentrate your attacks on Kethéric. He doesn't have that much health, and he should quickly fall to your blows, especially if you use spells. Once at 50% life, Kéthéric will trigger a cutscene and flee into the depths. You will have to go after him.

❖ Second fight in the Flayer Colony

After crossing the Illithid Nest, and fighting several battles, you will reach a flying platform leading to Kétheric's hiding place. Make sure you have freed Mizora before going there, otherwise an unfortunate fate awaits Wyll.

Having Gayle in your party will give you an additional option. Indeed, when you approach Kétheric and the other chosen ones, the Absolute will appear. You can choose to detonate the orb in Gayle's chest, which will kill everyone. The end credits will roll, and you are entitled to a short epilogue. It's fun to experience once, but it's not the real ending.

You can try to persuade Kétheric to abandon the fight and redeem himself, which will push him to suicide, like other bosses before him. But the developers at Larian have a sense of humor, since it will backfire on you this time. Indeed, if Chantenuit is alive, she is a prisoner again, which will confer immortality to Kétheric, then to the Avatar of Myrkyl who will replace him. There are also many enemies in the room. It is easier to face Kétheric, and tank him while you free Chantenuit, while killing all the other enemies.

When Ketheric dies, or commits suicide, he is replaced by the Avatar of Myrkul, a large, much more formidable monster. He slashes with his scythe and uses various necromantic spells. It can also suck the life of enemies still present. This is why it is better to clear the area before killing Kétheric.

The Apostle of Myrkul fortunately doesn't have much life, by focusing all your attacks on him, with a Speed   effect for example, you should be able to kill him in a turn or two.

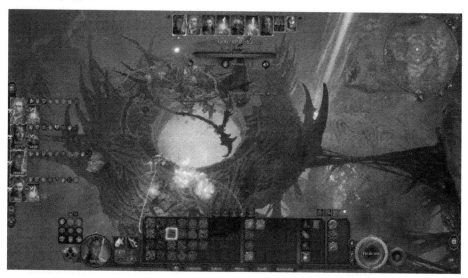

❖ Booty

Ketheric is well equipped, you can obtain his very rare armor, his hammer, and his shield. Without forgetting his infernal jewel, which you will need during the rest of the story.

Akabi and the Wheel of Fortune

The start of Chapter 3 of Baldur's Gate 3 has no shortage of interesting opportunities, when you know where to look. Akabi may be incredibly contemptuous and dishonest, but it's also one of the funniest and most profitable passages in the game.

To begin, you have to enter the circus area, which is not very difficult by talking to the doorman, or going through the gate at the back of the area.

❖ How to win at the wheel of fortune?

The important thing is to carefully select the character with whom you will speak to Akabi. Prefer someone with an excellent Perception score, like Shadowheart or Astarion, with the Assist spell, or even Wisdom of the Owl. Also save the game before turning the wheel. Indeed, you have only one opportunity to detect the trick used by Akabi, with each of your characters. Continuing to play afterwards has no point. If you fail the Perception roll, you must load the game, or play with another character hoping to have better luck.

To prevent Akabi from cheating, you have at least 2 possible solutions:

- The first is to sneak a Rogue like Astarion, behind Akabi's back, and use Pickpocketing. Steal his Jinn Ring, which he uses to cheat.

- The second method is to use a Bard, who has special dialogue options, to distract Akabi at the critical moment.

You then have to pay for a second turn of the wheel. If successful, you will win the jackpot. Ideally, this should also be a formidable character in combat, even solo. Like Lae'Zel or Karlach with a large stock of potions, or Astarion to cross the area discreetly. An Invisibility spell is also a good way to rush towards the portal while ignoring enemies.

Make fun of Akabi, who can't believe it. But as he is a bad loser, Akabi will teleport the big winner, alone, into a hostile jungle full of velociraptors. The objective is to reach the portal at the end of the area, but don't forget to search the nearby chest, which contains the big prize. You've indeed hit the jackpot: the legendary Nyrulna spear.

Take the portal then takes you back to the circus.

❖ Nyrulna

This legendary spear is much better than those available through Shadowheart quests. It inflicts +d16 Thunder damage, extends movement and jumps by 3 meters it provides immunity to falling damage, it has a gust power to dissipate cloud effects or push back enemies.

And above all, it has a formidable charge at very long distance, which will inflict heavy damage on all characters and enemies in its path. She is absolutely fantastic for rushing into the crowd at the start of a fight, which can instantly kill the most fragile enemies.

You'll hit a pack of enemies like thunder with it, which further extends your movement as a bonus! To get the most out of this spear, consider taking the Polemaster Feat.

Drooling the Clown

Visiting a circus in Baldur's Gate 3, which is more like a fair, wouldn't be complete without the typical activities: Akabi's rigged lottery, fortune-telling, a mass escape from killer animals, and of course, a clown who invites you on stage. Although it must be said that some unexpected twists were brought to the show.

❖ Start the quest for Drooling the Clown

You have to start by entering the circus zone, one way or the other. If you can't convince the gatekeeper, you can go through the back of the area, there is a gate you can take past the Temple of Illmater next to it.

As you approach the crowd watching the clown, he will start telling his jokes to which you can react in different ways. He will then invite you on stage. You are free to pass the role to one of your teammates, which can be particularly funny. Saving the game and testing different combinations is a good way to have a laugh. However, the result is always the same: the shapeshifters attack, and the false Drool must be killed.

Possible bug: If all the NPCs in the area remain stuck with a combat animation, even after the end of the encounter, go kill the eclipsing beast that you have locked in its cage.

Then, go talk to Lucretia, the necromancer who manages the skeleton dance, so that she can ask you to find the pieces of the real Drool.

❖ Find the pieces of Baveur the Clown

There are a total of 7 clown pieces to be found in town. This is by far the most difficult quest in the game when you do it blindly, without knowing where to look. There are usually traces of blood all over the area, or dead bodies, this helps to know when you are in the right place.

- Severed clown hand (X: -90, Y: -66): The first piece is not far away, it is on a plate behind the circus Kobolt. Steal it, or convince the Kobolt that you need it.

- Severed Clown Torso (X: -65, Y: -1025): On the body of a victim of the shapeshifting assassins, in the cave below the Temple of Illmater. You can go through a hidden trapdoor in the temple, which requires successful Perception checks, or through the entrance located along the river.

- Severed Clown Pelvis (X: 121, Y: 4): In one of the locked houses in the lower city. Unlock the blue door, north of the Basilisk Gate fast travel point.

- Severed Clown Arm (X: 39, Y: -95): In the basement of a house in Peartree in the Lower Town. You have to move a crate to reveal the trapdoor.

- Severed Clown Leg (X: -79, Y: -74): The basement of Lavernica's house. It's full of dead spiders. Find the trapdoor, then search the bodies in the circular room.

- Severed Clown Foot (X: -91, Y: -107): In the Rainforest house. You have to pick the front door or use the Misty Step while looking through a window.

- Severed Clown Head (X: -65, Y: 1041): On the body of Wilting Alex, in the ruins of the underground city. The entrance is in the upper left corner of the sewers. Be careful, this requires progress in Red Orin's quest to enter the zone, and this may have repercussions on one of your companions.

❖ How to manage the weight of clown pieces?

Even your strongest characters are likely to suffer from an overload problem if they are carrying all the Clown Pieces. It is advisable to send them to the camp chest with a right-click in your inventory. They will be automatically transferred when you speak to the quest giver to return them.

❖ Reward

You will receive the Gloves of Spell Power, which are reminiscent of the Heavy Slugger / Elite Archer Feats, but less well: they inflict a -5 penalty on the chance of hitting the spells concerned, in exchange for a bonus of 1d8 damage. They are mainly intended for practicing Occultists.

House of Hope

During Act 3 of Baldur's Gate 3, you quickly find yourself in contact with entities far beyond your mere mortal condition. Among the most nefarious beings standing in your way is Raphael, one of the demons you have already encountered throughout the previous acts. The narrative choices he offers are crucial for the rest of the game, and you will have to choose carefully how you address him.

○ *SPOILS*

This guide reveals some major elements of the plot of Baldur's Gate 3. We recommend that you read it on the sole condition that you are completely stuck in the game, otherwise you risk spoiling some major surprises in the game's story.

❖ How to get the Orphic Hammer in Baldur's Gate 3?

When you find Raphael in Act 3 of Baldur's Gate 3, he is on the upper floor of a brothel called Sharess' Caress. He is in his room soberly titled the Devil's Den. The demon then offers you a deal: he offers you the Orphic Hammer useful for freeing Orpheys, the prince of the Githyanki, and in exchange you bring him the Crown of Karsus.

Obviously, two choices are available to you:

1. You refuse, in which case you can always reconsider your decision even very late in the game, or try to acquire the hammer in another way described below in this article

2. You accept. You then obtain the Orphic Hammer, a legendary one-handed mace, and Raphael makes you sign his infernal contract.

If you have accepted, be aware that the terms of this type of contract are often rigged, so it is advisable to contact Helsik the diabolist so that she can advise you on the possibilities available to you. Obviously, there are quite few of them, as you probably expect.

❖ How to enter Raphael's Abode of Hope?

If you refuse Raphael's pact, or if you accept it but decide to free yourself from it, then things take a very different turn for you. You indeed have the possibility of going by your own means to Raphael's Abode of Hope, his infernal place of stay, even without his approval.

To achieve this, you will need to go to a shop called The Devil's Part north of the Lower Town of Baldur's Gate.

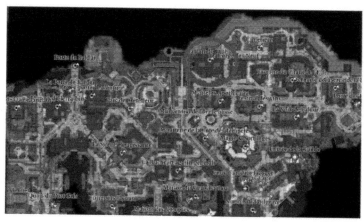

When you enter, you will see that a Dwarf is serving as a merchant at the counter. Go to her to have access to a relatively short but very important dialogue. It teaches you that you can go to Avernus on your own. The problem is that she demands a colossal sum of gold in exchange (20,000 gold coins).

Two possibilities are then available to you:

1. You accept the offer
2. You refuse the offer
3. You attack him and attempt to assassinate him

1. Accept Helsik's offer at The Devil's Part

If you accept Helsik's offer, congratulations, you've just been robbed of a good portion of your gold in exchange for a rather cordial relationship with the lady of the house. She then offers you a bag containing five objects, as well as a Grimoire bound in imp skin containing the instructions necessary for opening a portal towards Avernus.

All you have to do is go to the first floor of the house to participate in the ritual. Beware, at the top of the stairs there is an explosive trap. If you combine water with the imp head hanging on the wall to your right it will deactivate the trap.

All you have to do is place the items contained in the bag that Helsik gave you to open the portal, making sure to orient the camera so that the cardinal points (North, South, East, West) indicate that North points towards the room to your left:

- A Skull on the Western Point
- An Incense on the circle near the door opposite the Skull you just placed
- A Diamond or a Black Diamond on the circle above the one where you placed the Incense
- An infernal marble in the central pentagon

As soon as all the elements are placed correctly, the portal to the Abode of Hope will open, giving you free rein to dive into Raphael's infernal abode... Or not.

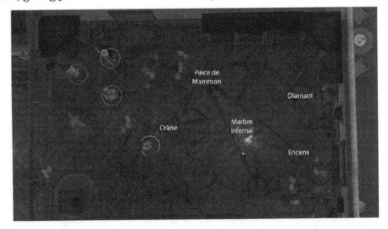

◇ *2. Refuse Helsik's offer to The Devil's Part*

If you choose to decline Helsik's offer, this is not permanent. You can always come back later and tell her that you have finally chosen to accept her deal, or try to kill her. But if this is your last word on this matter and you do not go further, then the Abode of Hope will remain inaccessible to you.

◇ *3. Attack Helsik and attempt to assassinate him*

The last option, the least prudent of all, is to attack Helsik head-on. If you do this, know that it immediately causes a horde of demons from Avernus to appear composed of

● Helsik: 96 life points (she also summons Golden Infernal Boars with 85 life points)a

● 2 Golden Minotaurs: 150 health points each

● 6 Golden Imps: 76 life points each

If you come out of this fight alive (it's possible, we did it), know that all is not lost. Indeed, even if your interlocutor has not told you how to get to Raphaël, on Helsik's corpse, behind the counter, you will find a Grimoire bound in imp skin as well as a ritual purse containing the objects you will need. needed for your stay in Hell. In the grimoire you will find a riddle detailing how to place different elements on a pentagram in order to open a portal to Avernus.

All you have to do is go to the first floor of the house to participate in the ritual. Beware, at the top of the stairs there is an explosive trap. If you combine water with the imp head hanging on the wall to your right it will deactivate the trap.

In Helsik's Chamber, place the following items on the pentagram painted with blood on the ground, making sure that the North of your mini-map points towards the room to your left (as you enter):

● A Skull on the Western Point

● An Incense on the circle near the door opposite the Skull you just placed

● A Diamond or a Black Diamond on the circle above the one where you placed the Incense

● A Mammon Coin in the circle above where you placed the Skull

● An infernal marble in the central pentagon

The portal should then open. Enjoy your stay in Avernus!

in Lorraine

Baldur's Gate 3 wouldn't be a worthy successor to the franchise without at least one evil Wizard experimenting in his tower. Lorroakan fills this unenviable role, and quests related to him can lead to radically different results. For the record, we already faced an evil magician in the Tower of Razamith in Baldur's Gate 1, he also wanted a woman of an exotic race to be delivered to him. The difference is that it was on the ground floor, and there was an ability increase book in the tower at the time. Someone must have read it, which deprived us of that opportunity in this sequel.

❖ Chapter 3: Magic & Witchcraft

After Gortash's coronation, you finally have access to the lower city of Baldur's Gate. You have many reasons to visit the city's most famous shop, the very recognizable Magic & Witchcraft shop, whose roof is covered in colorful stained glass windows. Gayle wants to visit the place in order to obtain a book on the crown of Karsus.

If he survived, you will also come across the mercenary who told you about Chantenuit at the entrance to the shop. This pretentious loser is asking for half of the reward for Chantenuit, since it was thanks to him that you learned about the bounty. Quite a few options are available in your responses, you can deceive him, persuade him, intimidate him, tell him the truth and more. This can lead to its reappearance a little later, or to its complete disappearance. But he is rather inconsistent in general.

Once in the shop, go upstairs and talk to the image of Lorroakan. Mention that you have information about Chantenuit. He will then open 4 portals. You must take the blue gate, on the far left. We have not tried to take the others, but we are sure that the result would not be positive.

❖ Dialogue with Lorroakan

You then meet this friendly magician, who obviously has nothing better to do than mistreat his assistants. There are many dialogue options available depending on Chantenuit's fate during Chapter 2, and your inventions.

You can start by asking him about his intentions, which will clarify things. He intends to lock Chantenuit in a magical cage in order to become immortal like Kéthéric. Here are the main options available to you:

- If you killed or delivered Chantenuit to Kéthéric, you can tell him the truth. Otherwise, it is also possible to deceive him and make him believe that Chantenuit has disappeared. But that doesn't do you much good.

- You can also confront Lorroakan there and kill him, in order to plunder his goods and his tower. It can be an uphill battle. You can loot his equipment and tower freely afterwards.

- It is possible to obtain some concessions from him by telling him about Chantenuit's position, or by promising to bring it back to him.

Then return to your camp. If you saved Chantenuit during the previous chapter, you can tell her that a magician and a shady mercenary have designs on her. You can then encourage him to go to the Tower of Razamith to resolve the problem together.

Then return to Lorroakan, Chantenuit will then appear with hostile intentions towards him. You only have two choices left at this point:

- Betray Chantenuit, to deliver her to the magician, which will earn you a reward of 5000 Gold, but which will seriously anger some people, including Isobelle. Lorroakan will assist you during the final battle with his tower.

- Help Chantenuit kill Lorroakan. A big battle ensues. In case of victory, Lorroakan is executed by Chantenuit, when she breaks his spine on her knee, ouch! It has some useful items for spellcasters. Chantenuit appreciates your assistance, and she will in turn help you during the final battle, with Isobelle.

If you saved the Tieflings during chapters 1 and 2, one of them will assist you during the fight, and he will then take Lorroakan's place. He will help you with the tower during the final battle. This is by far the best option.

In any case, remember to search the lower floors of the tower, as well as its underground passages. There are many great items here, including a legendary staff, books with permanent bonuses, and more.

Thrombus and the Jars of Karkass

The undead are all difficult to kill in Baldur's Gate 3, whether it is Kéthéric Thorm, Cazador or Karkass, there is not a single one who agrees to die normally. Fortunately, you are resourceful.

❖ Lift the curse of Oskar the painter: go see Karkass

- This isn't the only way to get in touch with Thrombus and Karkass' quests, but it is the only quest that sends you directly to speak to the latter. If you enter his house. You can remove rust from the side door, or go through a hole in one of the walls at the following coordinates: (X31, Y-166)

- You will need to adopt a smaller form to pass, such as a gnome disguise, a cat

127

transformation, or a potion of intangibility.

- Go talk to Karkass to ask for his help to lift Oskar's curse. He agrees to help you if you bring him the body of one of his zombie servants who fled: Thrombus. It is also possible to start this sequence of quests by meeting Thrombus by chance.

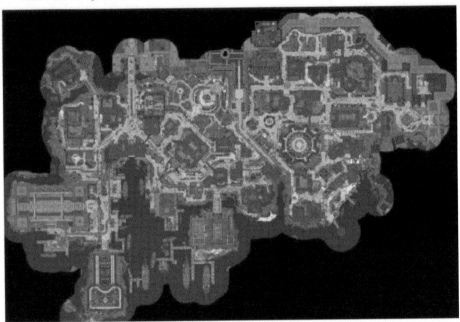

❖ *Where to find Thrombus?*

This zombie is hidden in a closet, in a house above Smugglers' Beach, a little to the southwest. The coordinates are (X57, Y-119).

Thrombus reveals to you that Karkass is an evil mummy lord (surprise!) and that it is possible to help Oskar without bending to his will by stealing the special torch in his possession, which requires killing him permanently preferably. You can choose to kill poor Thrombus in order to obtain his ring and bring him back to Karkass, he will then give you the required quest item: the Torch of Revocation. Or, you can go hunting for Karkass's magic jars, which grant him immortality. It's more effort, but it's much more satisfying and profitable.

❖ *Position of the 4 organ jars*

The first jar is in the morgue of the local cemetery, north of the city. Its entrance is located at the following coordinates: (X 27, Y 19). Just try not to be seen, it's a restricted area. Once inside, pick up the green jar on the table: Father Karkass's lungs.

- Go to the sewers now, follow the passage leading to the very north of the area. You will enter the old lair, lock the door which has a D4, but be careful, there is an ambush set by undead. Be careful, a special curse is present in the area, you

128

take damage when casting spells.

- The next room is a sort of zombie operating room. The brain jar is placed on the table, among the zombie pieces. The Liver Jar is in the nearby chest.

- Read the book "Form and Functions of Funerary Jars", which will give you clues about the position of the heart: it is in Thrombus. It's a love story really, he gave her his heart after all.

- Return to Thrombus with the Jars of Karkass, and he will give you the last jar upon request.

❖ *Affronter carcass*

All that remains is to destroy the jars, a simple method is to place them on the ground, then move away before firing a spell or a projectile to break them. This will trigger violent explosions. A poetic way to kill him is to sneak up on Karkass, set down the jars, and detonate them all at the same time as him.

You can now go and confront Karkass at his place, he is much weaker than in the past. Be careful with his spells though.

By killing Karkass, you will get his staff, excellent armor, scrolls and two keys. Use them to open doors and chests in the house. You will find the torch in the chest located at the following coordinates: (X:19, Y: -164). Thrombus will also give you a Crypt Lord Ring, allowing you to cast the level 6 spell: Creation of Undead.

Music

Orin may make you paranoid in Baldur's Gate 3, more than one player has found themselves attacking characters blindly in order to verify their identity. This is understandable, as there are also serious consequences for your group following the actions of this bloodthirsty murderess. It is possible to lose a companion permanently.

❖ Identity of the traitor in the camp

While attending Gortash's coronation, he will warn you that a traitor has infiltrated your camp, and that Orin can attack any of your companions. Unfortunately, there is not much that can be done in this area. She's not Yenna, even if the kid is shady. Refusing to let her into your camp won't protect you. Orin's kidnapping of someone is almost impossible to avoid.

❖ Save the companion taken hostage

The hostage-taking event has two triggers as far as we know:

- Take a long rest after Gortash's coronation, which will trigger a scene during which Lae'Zel claims to have discovered that Yenna is the traitor, and wants to execute her. You have to make an intimidation roll to save the girl's life. In reality, Orin has already kidnapped Lae'Zel, and adopted his appearance.

- The second trigger is to approach certain locations, such as the Emperor's hideout in the sewers (X:47, Y:979). This will make one of your companions not present in your group appear, covered in blood, who will announce Orin's attack. In reality, it was Orin who adopted his appearance, once again. However, there are more possibilities, and the abducted companion may be someone else. This includes: Gayle, Lae'Zel, Halsin, Minthara, and finally, Yenna, if the previous three targets are in your active group (Halsin and Minthara being mutually exclusive).

This means that by exploring the sewers with these companions in your party (Gayle, Lae'Zel, Halsin, or Minthara), you can ensure that Yenna is the hostage victim. It's much less debilitating than losing a companion. This also potentially allows you to ignore Orin's threats and go and confront him first. At least, if you don't mind sacrificing Yenna.

Theoretically, it is possible to reach Orin in his temple without triggering the hostage situation, by not resting and avoiding the relevant section of the sewers, but this risks causing some problems, given the importance of rest for the story progression and dialogue.

If the hostage situation takes place, Orin asks you to kill Gortash, and then go see her in her temple for a final confrontation. It is better to follow his instructions. If you decide to confront Orin before Gortash, the hostage will be executed.

❖ Enter the Forgotten City

Whether Gortash is dead or not, reaching Orin is always done the same way. You must gain access to the forgotten city, located under Baldur's Gate. To do this, you have to go through the Murder Court. But getting access to the murder court is already quite a long and complex ordeal. This requires completing the complex "Investigate the Murders" questline, with crime scenes all over Baldur's Gate. You must catch the assassin and kill him, in order to obtain his goods, including the key to access the Murder Court. We are not going to detail these quests here. To enter the court, you must go to the coffin maker, to the north of the city.

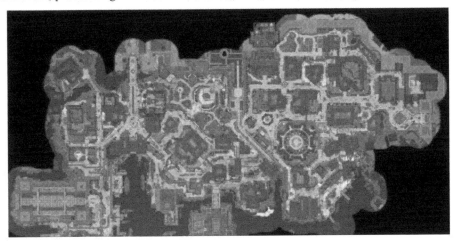

Pick his door. Inside, remove the painting with a butterfly on a skull to reveal a button. By activating it, a passage will open. You will need the key to enter.

The first obstacle in the court is a knight who will ask you for the hand of one of your victims. You can give him Gortash's hand if you have it, otherwise, you can give one of the hands of the murderer's victims. You can also kill him, if you do not wish to collaborate with this sinister prank.

In the next room, you will meet Sarevok himself. Once again, you can work together to become an unholy assassin, which will give you access to Orin's hideout but the price is high. You will need to kill an investigator, or Jaheira and/or Minsc, if they are in your group. You will gain fantastic light armor in exchange. A simpler and more positive option is to kill Sarevok and his gang, in order to obtain the medallion. This allows you to obtain excellent items.

You can now go to the entrance to the forgotten city, north of the sewers. You also have the means to open the door that was blocking your way.

.e door, and give one of the following answers if available:

. have Bhaal's locket.

- I killed Sarevok.

- I was recognized by the Murder Tribunal, I became an Unholy Assassin.

Several fights against assassins await you in the forgotten city, until you reach the depths of the Temple where Orin and the hostage are located.

❖ Fight against Orin

This confrontation is not particularly difficult at this point in the game. Orin transforms into a formidable pest in melee, but he is also too fragile to be a real threat by concentrating your attacks on him. Ironically, the other assassins on the platform are more troublesome, since they use a combination of invisibility and Sanctuary. Use spells and area attacks to kill them. The Gale and Thundering Wave spells are particularly effective, since they are located on the edges of a platform surrounded by void. This will teach them that being smart in such a poorly designed temple has a price!

❖ Orin and Dark Urges (spoilers)

Warning, major spoilers follow: Orin has very special ties to your protagonist if you choose the Dark Impulsions origin. Indeed, as you will discover along the way, you are a offspring of Bhaal, just like her. And your opposition takes a more personal turn, since you will both seek to be recognized as Bhaal's chosen one. It is possible to transform into a Ravager in his place, during the confrontation in the temple. Once Orin dies, you are finally recognized as the only chosen one, or you can reject your lineage.

Sewers

While the sewers in Baldur's Gate 1 were incredibly orderly, with a coherent structure, those in Baldur's Gate 3 are completely chaotic. Its different sections seem to date from different eras, parts have collapsed and been patched up, and everyone has created their own little private access. There must be so much traffic down there that you're more likely to run into people you know there than on the surface.

❖ How to enter the sewers?

There are so many passages leading to the sewers, that you should enter the area sooner or later, even by accident. However, here is a non-exhaustive list of entries, but which should not be far from it:

- The different manholes present in the lower city of Baldur's Gate. It's pretty logical. We counted at least three, including one right next to the Basilisk gate.

- Entering the Elf's Song tavern, descending into the cellar, then revealing the passage leading to the Emperor's hideout. At the bottom of the latter, a ladder leads to the sewers.

- The Fkymm warehouse has a passage leading to the sewers, although you will rather go the opposite way: enter the area via the latter.

- The back door from the bar in the Thieves' Guild leads to the sewers too.

- By digging the graves in the cemetery, you will eventually open a hole leading to the sewers.

- In the prison, which is practical for escaping, or organizing an escape.

- After recruiting Minsk, the exit from the reservoir becomes access to the sewers, but you cannot take it before, so this passage is useless.

- Finally, the exit from Cazador's dungeon leads to the sewers.

❖ Annotated map of sewers

You can consult an interactive map to have all the information on the area. Here is a simplified version with the most notable elements.

- Access to the submarine is fundamental to solving the quest for the Temple of Umberlee, and saving the Gortash hostages.

- The entrance to the reservoir requires you to solve a puzzle, you meet Minsc in the next area if you have progressed in his quest with Jaheira.

- The Ancient Lair is linked to the quest for Karkass and Thrombus.

- The entrance to the ancient city leads to Orin, it is also a mandatory crossing point to complete Act 3 and move on to the final battle.

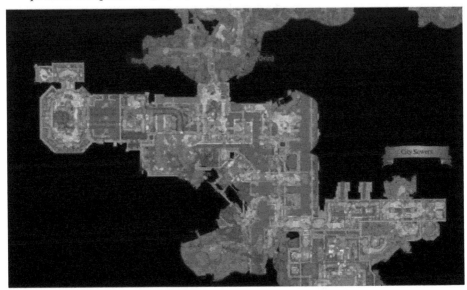

❖ How to enter Cazador Castle via the sewers?

If you've been paying attention, you must have noticed that there are some strange ornate barriers in the sewers, which lead to a door directly under Cazador the Vampire's castle. You can cross the barrier via a teleport, or by taking the raft along the waterway. You will even meet a woman invited to the "party" in the castle. Please note that it is impossible to pass through this door at the moment. You must enter the Cazador dungeon, and use this passage as an exit door, to go and rest, or to evacuate. It then becomes possible to take it in the other direction. It's a door to "Dark Souls".

❖ How to open the reservoir lock?

Navigating the sewers is a bit of a chore at times, but it's not too difficult. The only notable obstacle is linked to the puzzle allowing you to open the airlock leading to Minsc. It is necessary to equalize the pressure of the valves in order to open the door leading to the tank. The principle is simple, but you have to experiment a little before finding the right timing. Separate a member of your group to limit damage, then click on the left valve (water level) and a few seconds later, on the temperature valve. The goal is for both gauges to have the needle in the center at the same time when you activate the center valve.

You can click on the corresponding valves during the procedure to stop or restart the power, to help you balance the flow. It will probably take a few tries before you get the timing right. If successful, the sans below will open. Don't make the same mistake as us, and check regularly if it has opened or not. We didn't have any alerts in our game. Possibly because of a bug.

Hunter

After hearing about Astarion's cruel master during two chapters of Baldur's Gate 3, the time has come to confront him. That is, if you can find the entrance to his castle We can say that he is well protected from invaders, since the main entrance to his home is directly part of the content cut from the game. While waiting for a possible addition in the future, you will have to go through the back door.

Before going into the castle, you should talk to the Gurrs in the refugee camp, before the bridge leading into town. They are vampire hunters and Cazador kidnapped their children (via Astarion). Promising to try to save their children can have some interesting repercussions down the line.

❖ How to enter Cazador Castle?

This is completely unintuitive, but you have to climb the guard tower in the center of the lower city of Baldur's Gate at the following coordinates X:-68; Y:-53. It is located opposite the Magic & Sorceries store. There is also a teleport point nearby. Climb to the top to reach the top of the city walls. Next, head to the north of the area, the back of Cazador Castle can be seen on the map. Along the way, you will come across a group of guards in a second tower. You can bully them, or kill them. Then all you have to do is enter the castle.

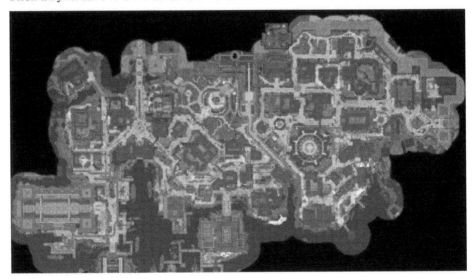

Let's mention in passing that there is also a door in the city sewers, leading to the depths of the castle. But it will serve as a shortcut later. It cannot be opened from this side.

❖ How to open the sinister door?

Once inside the castle, a strange ornate door will block your way. You need two items in order to open it:

- Open the door to one of the rooms, pass the corpse with the necrotic aura and search the cupboard at the back of the room to find the Kozakuran dictionary.

- Go down the stairs until you find a secret passage leading to the kennel (Perception check required). Inside, you will come across an undead named Godey. Kill him and take the Szarr family ring.

135

Prepare for a big fight against a pack of werewolves after opening the door. Next, head to the elevator to descend into the depths of the dungeon.

❖ What to do with the vampires imprisoned by Cazador?

Cazador's poor victims were all transformed into vampires. Unfortunately, there is nothing you can do for them. It is possible to release them, but this will have negative repercussions. It's best to leave them all in the cage for now, and decide their fate later, after defeating Cazador. Of course, you can also release them to appreciate the tragedies that will ensue.

❖ Fight against Cazador and interruption of the ritual

Before entering the large ritual room and confronting Cazador, it is advisable to search the dungeon carefully. Finding the skull of Cazador's master, and learning the secrets of the ritual will give you additional options for the future. Also remember to summon reinforcement creatures and rest.

Once faced with Cazador, the dialogue begins. If Astarion is in the group (which we strongly advise you to do), he will be taken prisoner by the sacrificial altar. It is also possible that it is already present, if it was kidnapped by the vampires. The fight against Cazador begins, you have 7 turns to kill him, otherwise the ritual will end and all the prisoners will be sacrificed, including Astarion. Cazador will also become much more powerful.

Send someone to free Astarion urgently, use the basic ability "Help" to free him, which will allow him to take part in the fight. Not foggy or Speed   allows you to reach it quickly. It is strongly recommended to concentrate your entire group on Cazador in order to kill him as quickly as possible. Like all vampires, he is extremely vulnerable to sunlight and radiant damage. The solar light spell inflicts 20 damage at the start of the turn. The Blood of Lathander is an ideal weapon for this fight. Spells like Solar Beam and others are devastating when used on him.

136

- Don't waste time killing the bats, reinforcements arrive at the end of each round.

- Killing the sacrificed vampires is a good way to weaken Cazador, but it will deprive you of several options later.

- When Cazador has been killed, finish off his minions, before interacting with the coffin in the middle of the area.

❖ Awards

Cazador holds a bank safe key, and a very rare magic staff: Misfortune. It is perfect for a Necromancer, and for any user of control spells.

❖ Astarion and the ritual

Finally, after defeating Cazador, and extracting him from his coffin, the time for choices came. There are quite a few options available and variations.

The best choice, if you decide to play someone virtuous, is to simply let Astarion kill Cazador. Tell him to forgo the ritual, because it's not like him, he's better. Then, we must decide the fate of the other vampires, whether the offspring of Cazador, or the thousands of prisoners (including children). This obviously depends on your preferences and ethics. Killing all the prisoners in order to free them from their suffering, and letting the offspring go into exile in the Underdark seemed to us to be the best choice. You can also choose to kill them. As you leave the dungeon, you will come across the Gurrs. Tell them the truth about their children, and they will agree that you did the right thing. They will even side with you for the final battle.

If you found the skull of Cazador's master and passed the 3 Wisdom checks while holding it, you learn how to complete the ritual. If all of Cazador's spawn are alive, then you can continue the ritual. Astarion will need your help to copy the symbols on his back, and engrave them on Cazador's.

He will then begin the ritual. At this point, it's too late to change your mind. If you force Astarion to interrupt the ritual, he will leave the party permanently and attack you. By completing the ritual, Astarion will become the ascendant vampire. Its bite becomes much more powerful, and its attacks deal Necrotic damage. He also becomes much more evil. As you leave the dungeon, you will come across the Gurrs, but this time the fight will begin.

Cave of Magic and Spells

One of the first things to do once in Baldur's Gate is to go shopping at the Magic & Sortilèges store in the lower town. As the building is massive, and its appearance is unique, it is difficult to miss. But his vendors are just the tip of the iceberg in Baldur's Gate 3, and he hides several secrets.

❖ How to enter the cellar storing banned books?

There are two methods to reach this underground area. The first is to talk to the librarian at the back of the store, Tolna the bookseller, and ask her questions about the books, in the company of Gayle. By using persuasion, she will give you several clues on how to use the portal. Go upstairs, then unlock the door on the left. You will have to be discreet, since it is a restricted area and a guard has his eyes trained on the entrance. Using a Major Invisibility spell is a good idea.

Search the room and pick up the key. Then interact with the portal to enter the cellar.

The second method does not require discretion, but it is more complex. You must enter the Tower of Razamith to speak to Lorraakan. To do this, announce to its mirror image that you have information on Chantenuit. Then, take the left portal (blue).

Once in the tower, go down to the lower floors, by jumping on flying objects, or using teleportation spells. Inspect the different magic buttons accompanied by a small panel. You have to use the one that leads to the Cave. It will make you appear directly in a room containing one of the latter's most precious books.

❖ Cellar map

There are two sections in the cellar. The first is quite conventional, and it leads to the Book of the Red Knight's Final Stratagem. Reading it will give you a unique spell scroll, which Gayle can copy. This is also where you will start, if you go through Razimith's tower.

The second section of the cellar is probably the one that gives you trouble, with its successive series of doors. It's actually quite simple, even without a guide, if you go about it methodically. Having a Rogue or Bard with a high score in Perception and Disarming Traps will make your life much easier.

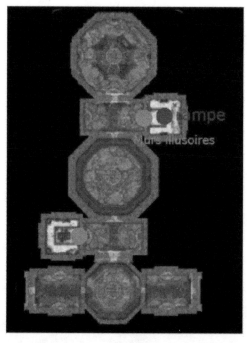

❖ How to solve the riddle of the doors?

The solution is to take the correct series of doors each time, until you reach a lever to activate. You must separate this character from the rest of the group, and send him to explore the area alone in order to limit the damage received, and avoid triggering traps by accident. You can bypass traps, or enter doors before the magical effects reach you. Here is the precise route to follow:

● Silver Hand > Evocation > Wish > Activate the Elminster Gate Lever

Return to the first room of the maze.

● Silver Hand > Abjuration > Silver > Activate the Karsus Gate Lever

You can now search the rooms behind the door of Elminster and Karsus. They are full of illusions and treasures, so remember to save the game and watch your perception rolls.

Behind Elminster's door you will find the Codex of the Triarch, which gives a temporary hit point bonus, as well as the ability to finish reading The Necromancy of Thay. Which will unlock a powerful undead summon.

Behind Karsus' door are the Annals of Karsus, for Gayle, which will trigger dialogues with the latter, if he is present.

❖ How to get out of the Djinn Lamp?

There are two illusory walls in the area, you can pass through them by walking inside them, much like in Auntie Ethel's Overgrown Tunnel. Check out the location map above.

One of the illusory walls hides a magic lamp. By interacting with it, it is possible to find yourself trapped in the lamp. Getting out of it is simple: use any summoning skill. Whether it's Scratch, a Zombie, a Familiar, etc. There's also a summon scroll in the lamp, if you don't have anything else.

Tower of Razamith

You may not know it, but Razamith's Tower was already explored in Baldur's Gate 1, and navigating inside was already a pain. This is still the case in Baldur's Gate 3, especially since it has become significantly more complex. This is a fairly important area, linked to Gayle's personal quests, as well as to its owner, Lorroakan, who intends to get his hands on Chantenuit.

❖ How to enter Ramazith Tower?

It's a bit silly, but with the content cut from Baldur's Gate 3, the upper city is not present, so the same goes for the physical entrance to the tower. Fortunately, the developers have found a trick. After Gortash's coronation, you finally have access to the lower city of Baldur's Gate. You have many reasons to visit the city's most famous shop, the very recognizable Magic & Witchcraft shop, whose roof is covered in colorful stained glass windows. Gayle wants to visit the place in order to obtain a book on the crown of Karsus from his cellar.

Once in the shop, go upstairs and talk to the image of Lorroakan. Mention that you have information about Chantenuit. He will then open 4 portals. You must take the blue gate, on the far left. We have not tried to take the others, but we are sure that the result would not be positive.

Once inside Razamith's Tower, you can speak to Lorroakan and decide how to conduct your transaction with him. We suggest you return with Chantenuit to kill him in order to plunder his goods quietly.

❖ How to descend the Tower of Razamith?

Magicians being dirty elitists in this universe too, there are no stairs in the tower. People who are unable to use magic are also considered disabled in their eyes, and they are clearly not willing to put up a ramp.

You have several methods of getting down to the lower floors: the one available to everyone is to jump on the flying furniture in the middle of the tower. Imagine you are playing a platform game. Jump from table to table until you reach the lower floor.

A significantly faster method is to use Fly, Misty Step, or a fall damage reduction effect, to get down in one go. You can also do it from outside the tower, through the windows.

Once on the middle floor of the tower, you will encounter monsters, traps, arcane turrets (use lightning damage) and especially Frame Buttons. Check the small plate on the front to find out their function. Some will activate the tower's security measures.

Two buttons in the frame in particular interest us: "Down" in the northwest corner which allows you to reach the lower level on which the best treasures are stored, and "Cave" leading into the Magic & Sorcery Cellar. Some plates are invisible, plan a potion, spell, or other effect to read their description.

❖ How to deactivate the arcane barrier?

In order to access the two treasures protected by the spheres, there is a small puzzle to solve. Once again, you need to see invisibility. Use a spell, a potion, or have Volo violently operate on you. When you have detected the two levers, you must succeed on an Arcane D20 in order to activate them. Gayle with the Assist spell is by far the most proficient in this area. If you fail, a trap will be triggered.

When the levers have been used, the spheres will disappear, and you will be able to collect your precious items, a legendary staff and the Robe of the Weave.

Temple de Baal

The final battle of Baldur's Gate 1 pitted you against Sarevok in the Temple of Bhaal. Baldur's Gate 3 multiplies the references and links between the two games, and if it is not the final battle in this case, it will serve as the conclusion to chapter 3 for many players. The temple also appears in the main menu of the game, which clearly shows its symbolic importance.

❖ How to enter the Bhaal Temple?

For once, there is only one possible access, with an impassable door. The three methods of forcing it open are closely related. To do this, you have to go through the Murder Court. But getting access to the murder court is already quite a long and complex ordeal in itself. This requires completing the complex "Investigate the Murders" questline, with crime scenes all over Baldur's Gate. You must catch the assassin and kill him, in order to obtain his goods, including the password to enter Murder Court. You can also kill the victims on the list yourself, in order to receive the password, but expect repercussions from your companions and the guard. We are not going to detail these quests here. To enter the court, you must go to the coffin maker, to the north of the city.

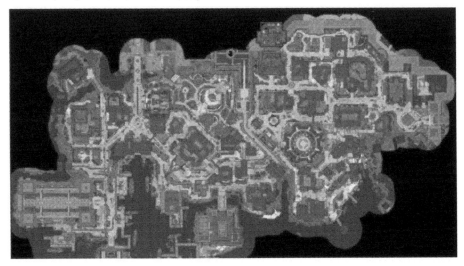

Pick his door using discretion, or use the key if you have it. Inside, remove the painting with a butterfly on a skull to reveal a button. By activating it, a passage will open. You will need the password to be able to enter.

The first obstacle in the court is a tombstone knight who will ask for the hand of one of your victims. You can give him Gortash's hand if you have it, otherwise you can give one of the murder victims' hands. You can also kill him, if you do not wish to collaborate with this sinister prank.

In the next room, you will meet Sarevok himself. Once again, you can work together to become an unholy assassin, which will give you access to Orin's hideout but the price is high. You will need to kill an investigator, or Jaheira and/or Minsc, if they are in your group. You will gain fantastic light armor in exchange. A simpler and more heroic option is to kill Sarevok and his gang, in order to obtain the medallion. This also allows you to obtain excellent items.

You can now go to the entrance to the forgotten city, north of the sewers. You also have the means to open the door that was blocking your way.

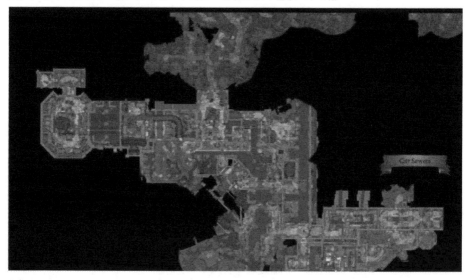

Click on the door, and give one of the following answers if available:

● I have Bhaal's locket.

● I killed Sarevok.

● I was recognized by the Murder Tribunal, I became an Unholy Assassin.

❖ Ruins of the Undercity

Several fights against assassins await you in the forgotten city, until you reach the depths of the Temple where Orin and the hostage are located. The most dangerous is certainly the Murder Inquisitor, who ambushes you at the stairs. He has assassins and crossbowmen positioned around the area. He will apply a special effect to your characters, who will die if you don't kill him quickly enough. You have to hunt him down aggressively, in order to kill him in time.

❖ Save the hostage

Be careful, if you enter the Temple of Bhaal before killing Gortash, the hostage will be executed and he cannot be brought back to life. It's not too bad if it's Yenna (monster), but it's a lot less fun if it's one of your companions, which is much more likely to happen. So it is advisable to kill Gortash first. The way you pass through the city gate has an impact on the dialogue with Orin, she will blame you if you killed Sarevok in her place, but that's it.

❖ Fight against Orin in the Temple of Bhaal

After making your way to the temple, you will discover that the cultists inside are not hostile. They are there to observe your fight against Orin. You are obviously free to kill them, if you want to rid the city of these vermin, or quench your bloodlust before the main course: Orin.

This encounter shouldn't normally be a problem if you play normally. Orin transforms into a formidable melee pest, but he is also too fragile to be a real threat by focusing your attacks on him. Ironically, the other assassins on the platform are more troublesome, since they use a combination of invisibility and Sanctuary. Use spells and area attacks to kill them. The Gale and Thundering Wave spells are particularly effective, since they are located on the edges of a platform surrounded by void. As they don't have any notable objects in their possession, have a good laugh while showing them the Russian method of execution.

❖ Dark impulses

Warning, major spoilers will follow. As you probably know by this point, if you play Dark Urges, you are a spawn of Bhaal, and in the past, you fought Orin in order to become his chosen. The confrontation against Orin is therefore personal, it is the return match. It is possible to transform into a Ravager in his place, during the confrontation in the temple. Once Orin is dead, Bhaal will speak to you, so that you become his chosen one. There are two possibilities :

● Accept Bhaal's gift, become his new chosen one and the pest.

● Reject your lineage, which will cause your death. But Blight will bring you back to life, as a true hero freed from his origins. Let us note in passing that if she is alive, Minthara risks leaving you for having wasted your potential.

ALL ABOUT CLASSES

Tier list des classes

The idea of   proposing a Tier List of Baldur's Gate 3 classes made us want to tear our hair out, but we must recognize that all classes are far from equal, we could say that some are more RP than others. others. While on a tabletop, a competent game master can tailor his campaign to the classes his players select, this is absolutely not the case in a video game, which typically includes much more combat and very different balancing. In addition, Larian Studios has rebalanced (or unbalanced) quite a few elements. The level 12 cap also has a notable impact on everything. Note that multiclassing is also an important element of the game.

❖ Go on an adventure your own way

Of course, you should not feel obliged to follow the classification below to the letter To begin with, it is obviously subjective, not all players give the same importance to each element provided by the classes. The most important thing is to play the class you want. What really makes Baldur's Gate 3 so charming is the ability to approach the adventure in a certain way, whether diplomatic, deceitful, benevolent, or simply linked to the mechanics of a particular class. In a sense, you can see the Tier List below as suggestions for possible approaches to gameplay.

❖ Scoring criteria for this Tier List

The two main criteria in the ranking below are:

1. Combat effectiveness, in one or more roles. Taking into account things like physical and magical DPS, controls, survivability, mobility, support, and versatility. If a person is only OP in a specific situation, it's not great.

2. The contributions of the class outside of combat. This can be during dialogues, or all ancillary activities, such as burglarizing all the places you come across.

3. To a much lesser extent, the potential of the class in the context of multiclassing also influences our choice. If a class is so good that you want to take a level or two of it in your build, it's usually not for nothing.

❖ Best classes in Baldur's Gate 3: S Tier

* Cleric

We don't say that because Shadowheart is really cute. Its stats and divine domain are horrible in combat, which can give a bad impression of the class before a respec it urgently needs. The Cleric is an incredibly versatile and powerful class, which can play almost any role in combat: Tank, melee DPS, magic DPS, Support, Healer depending on the Divine Domain selected. He's not necessarily going to be the best in every role, far from it, but when you put everything together as a class, the result is fantastic.

This will come as a surprise, but Magicians and Sorcerers are often tempted to take one or two levels of Cleric in their build, for the many benefits that this brings them handling of intermediate or even heavy armor, the shield, more spells in their repertoire and a divine domain that can create interesting synergies.

+ **Can do almost anything**

+ **Excellent non-combat bonus with Assistance**

+ **Overall robust**

+ **Perfect for managing Perception**

- **Big lack of Charisma**

- **Little variation in sources of damage**

- **Moderate physical damage**

  ◦ *Bard*

The Bard is the ultimate Jack of all trades, much like the Cleric. He can be competent in melee, especially with the Valor or Sword schools, while being both an offensive spellcaster and a decent healer, even if he is not the equal of any specialist class. It has many useful or formidable exclusive spells. As a bonus, he will inspire his companions to offer them significant combat bonuses and their skills. And if you choose to add 2 levels of Paladin to your Bard, then it becomes a real melee killer, who will distribute Divine Punishments in large numbers. But that's not all, it's truly the ultimate Skill class. It is not an exaggeration to say that the Bard can manage the majority of the group's rolls on his own, being almost unequal each time: Persuasion, Deception, Discretion, Sleight of hand, it is clearly the face of the group and its Dodger. It can also compensate for the absence of a Magician to cover its areas of predilection based on Intelligence.

The class has a mind-blowing number of exclusive dialogue choices, which are both useful and funny. It's like Jaskier from The Witcher became a competent protagonist while remaining petulant. The Bard's Performance is the solution to almost all dialogues, and you can even play music in public (with real music), which will attract NPCs in the area, who will congratulate you and throw coins at you Golden. Even in terms of role play, the Bard is surely one of the best classes in the game.

+ **Can do almost anything in combat**

+ **Can handle almost any skill**

+ **Fantastic support**

- **Serves as the face of the group**

Damage that leaves a little to be desired unless you really optimize

❖ A Tier

Classes in this category are absolutely excellent, they just tend to be a little more specialized, or have a weakness. In our eyes, the majority of classes in the game fall into this category, which tends to show that a pretty good balance has been struck, and that it's difficult to find truly bad classes.

◦ Paladin

Our knight of good clearly deserves his place at the top of the rankings. Halfway between the Warrior and the Cleric, he is really powerful in melee, while offering some really useful holy spells. He is a very robust character, who tanks, DPS, and heals, to a certain extent. He also has the Charisma required to serve as the face of the group, in order to lead the dialogues. It is also a very popular class for multiclassing, it must be said that its first two levels provide so much, that even Clerics and Bards want to have them too. The only reason why the Paladin is not S Tier is that you will have to strictly respect his Divine Oath, which will greatly limit your actions, and severely punish unexpected repercussions, which can frustrate players who want to keep a specific subclass like Vengeance or Devotion. On the other hand, if you don't mind falling to the dark side and changing certain powers, you can then play a Forsworn, which also has its charm.

+ **Significant melee DPS**

+ **Robust**

+ **Serves as the face of the group**

- **Very low DPS at range**

- **The need to respect one's oath (or not)**

◦ Warrior

The Dungeons & Dragons 5 Warrior is excellent, it really is the martial class par excellence. She is very versatile when it comes to dealing damage, one-handed, two-handed, two-weaponized, or even ranged where she can outplay the Ranger. She can also wear any armor and use shields. It can even add some magic, or special tactical tools. This class is entitled to more gifts and bonuses than the others, and it is entitled to two powerful tools available at low level, with Second Wind to heal itself, and Fierce to trigger additional attacks. This becomes even more powerful when he gains additional attacks. This explains why so many classes like to multiclass into Warrior.

+ **Fantastic DPS in melee and/or ranged**

+ **Easy to play**

+ **Robust**

- **Very little contribution outside of combat**

◦ *Dodger*

A good group of adventurers necessarily needs a good thief, you need someone to scout the terrain, disarm the traps, open the locks, and to rob the merchants if you are unscrupulous. Suffice to say that he is one of the most active characters in the group outside of combat. He can even serve as the face of the group, with Deception and Persuasion, and a decent Charisma score. But it's also an incredible asset in combat, even potentially the biggest DPS of the group, whether at a distance or in melee, thanks to its sneak attacks. By choosing the Rogue Assassin subclass in particular, it is possible to instantly kill isolated enemies, or engage in combat with a heavily injured target. He can then plunge back into the darkness at the first opportunity to start again.

+ Devastating sneak attacks

+ Perfect for those who like infiltration

+ Viable in melee and at a distance

+ Essential skills outside of combat

+ Can be used as an extra face

- Remains very fragile

- You have to have the patience to play it well

◦ *Magician*

Like the Rogue, the Warrior and the Cleric, the Magician is one of the iconic classes of the base group. It's also the only class in the game not to have 8 Intelligence, he must be depressed being surrounded by idiots (this also explains Gayle's somewhat condescending attitude). This has the advantage of offering advanced mastery of many otherwise neglected skills such as Arcana, Investigation and History. The Wizard is an arcane spellcaster, with the largest repertoire of spells in the game, giving him unparalleled versatility in this area, as long as you take the time to select them before combat (or after a loadout) . This allows it to also be a real support, since the Sorcerer and the Occultist do not have the leisure to sacrifice their meager selection of spells in this way. The Magician's many specializations can influence his gameplay, but we will especially remember the School of Evocation, which allows you to throw Fireballs and Cones of Frost, to kill all enemies at once without having to worry about your party being automatically spared from the AoE. Of course, all this has a very high price, the Magician is terribly fragile, and once his spells are exhausted, he is no longer of much use.

+ Widest variety of spells in the game

+ Excellent area magic DPS

+ Mass control spells

+ Utility spells with varied uses

+ Only Intelligence class in the game, and the skills that go with it

- Really not cut out to be the face of the group

- Very fragile

- Not very useful without his spells

  ◦ *Sorcerer*

The Magician's little brother is not very intelligent, but he is full of charm. The Sorcerer only has a small handful of spells in his arsenal, but he can cast a few more, and most importantly, he can modify their effect depending on the circumstances with metamagic. This forces Sorcerers to specialize and limit themselves to well-chosen spells, but it makes them the most devastating spellcasters in the game. Their specializations also go in this direction, with different exotic gameplay, such as unpredictable wild magic , dragonblood which will amplify their damage as much as possible, or storm sorcery, which offers mobility and automatic magical attacks on enemies when wielding lightning or thunder. At the same time, the Sorcerer can also serve as the face of the group with his very high Charisma. On the other hand, like its big brother, it is a very fragile class by default.

+ Devastating spells in large quantities

+ Attractive subclasses

+ Metamagics offering tactical opportunities

+ High charisma to serve as the face of the group

- A very limited range of spells

- Very fragile

- Not very useful without his spells

  ◦ *Occultist*

The junior class of spellcasters is the Occultist. By making a pact with a powerful otherworldly entity, he obtains spells and mysterious powers. He has a limited choice of spells, which he can only cast a handful of times per day, but in return, he undoubtedly has the best minor spell in the game: Eldritch Explosion. After a few levels, it becomes a formidable spell that he can cast without risking running out. The pact's bonuses amplify his powers, and can also give him other tools, such as pets, or the ability to be a more than honorable melee fighter, since he is a little more robust than his colleagues. Unfortunately, the very popular Hexblade build is not available in Baldur's Gate 3, we will have to wait for DLC for the class to show its full potential. Oddly, as this class is also based on Charisma, it can also serve as the face of the group.

+ The best cantrip in the game, by far, infinitely usable

+ Has notable exclusive spells

+ Viable in melee with certain builds

+ **Can serve as the face of the group**

+ **Wide variety of builds and gameplay**

- **Very limited range of spells and dependent on the overworld boss**

- **Few spells available each day**

- **Remains a little fragile**

  ◦ *Barbaric*

While we are used to seeing this class of big brutes in Diablo, we have only rarely had the opportunity to use it in Baldur's Gate and CRPPs of the same family. She does not disappoint in this 3rd installment, and Karlach is an excellent way to discover the charms she has in store. To begin with, it's a fantastic class for a very distinct approach to dialogues, if you want to solve all your problems based on howling, big muscles and angry blows, you will have fun, the options exclusive to Barbarian are excellent. In combat, this class is also fantastic, once enraged, it has exceptional melee DPS, with additional attacks, and the possibility of automatically having advantage on its targets, which makes it much easier to hit bosses and other stubborn enemies, even if it is at the expense of his own survival. Fortunately, it's the class with the most life in the game. The Barbarian also has gigantic leaps, even the ability to grab enemies and throw them into the void, or onto their colleagues. Needless to say, we have a good laugh. The downside of all this is that the Barbarian is really not made to cast spells, since he cannot cast them, nor maintain his concentration when enraged. It is nevertheless a rather popular class among fans of multiclassing, for its combat bonuses without armor and/or rage.

+ **Titanic melee DPS**

+ **The biggest life bar in the game**

+ **Exotic options with jumping and throwing enemies**

+ **Almost useless outside of combat**

+ **Very special dialogues**

- **Incompatible with the use of spells and their maintenance**

- **Recklessness can be deadly**

❖ *B Tier*

Finally, here are the classes that we consider to be a little worse than the others. They remain completely viable, and there is no doubt that they can even be fantastic with the right build and in the right hands. But it's not for everyone, it requires a lot of knowledge of the game and its mechanics, and a little more effort. This shouldn't stop you if you intend to play them.

  ◦ *Lurker*

The famous class of Drizzt and Minsc is not really made for a video game, since its bonuses are normally based on the environment in which it finds itself, and its enemies. The developers had to improvise more versatile compromises, which are not without interest, but which are far from matching the tools available to other classes. The Ranger can be a good melee fighter, and especially at range, but the Warrior surpasses him in both areas. It can also replace the Dodger, in terms of skills, as well as with more stealth-based gameplay, but without the sneak attacks. Versatile classes are at the top of the rankings, but the Ranger struggles to fit into this category a little. He can still find his place in the group and be a notable asset. We will especially appreciate it for the possibility of emulating Drizzt in melee, and Legolas at a distance.

+ **A good ranged DPS class**

+ **Quite robust**

+**Skills similar to Rogue if you sacrifice other bonuses**

+ **Some bonus spells**

- **Not the best in melee or even at range**

- **The class theme sacrificed**

  ◦ *Druid*

Even though the Druid's hairy sex antics served as publicity for Baldur's Gate 3, the beast's performance is a bit underwhelming the rest of the time. Halfway between a caster of sacred spells and a melee fighter, with metamorphosis, the Druid is not as convincing as one would like, especially at high level. Animal shapeshifters aren't very effective in combat compared to martial classes, but you can still perform unique actions, like going through small tunnels to infiltrate an area, or disrupting NPCs. The Druid's spells are quite useful, in addition to the different categories of usual spells, it is even possible to communicate with plants in addition to animals. Spells linked to the weather and modification of the environment are obviously absent, since it is very difficult to transcribe that in a video game. As with the Ranger, this makes the class a bit disappointing. Being mean, we could say that the Druid is a bad eco-friendly variant of the Cleric. In terms of skills, it covers those related to Wisdom, such as Perception, which is very important, but also Nature and Survival, whose impact is negligible. Let's still mention the subclass based on spores, which allows you to summon zombies. She's particularly amusing, but she didn't impress us.

+ **Unique options with out-of-combat transfigurations**

+ **Unique spells and natural abilities**

+ **Ideal for a good Perception score**

+ **Combat mix of spells and transformations**

+ **The Assist spell**

- **The combat effectiveness of Transfigurations is disappointing**

- **Devoid of Charisma**

◦ *Monk*

Finally, here is the class that likes to fight with its fists rather than with equipment, the Monk. The class is generally considered one of the weakest in Dungeons & Dragons 5. It was buffed a bit in Baldur's Gate 3, and it is now considered balanced from the perspective of many players, and we are pretty d 'agreement. It is possible to create an OP build with the Monk, but it is not within everyone's reach. This is why it is in our B Tier, and not in a lower Tier. By knowing what you're doing, the Monk can be an effective class, capable of unique and surprising actions not found elsewhere. The Monk uses his extreme mobility and veritable deluges of blows to annihilate his enemies, while defending himself with the sole skill of his body, his mind, and ki. If you don't like to rely too much on equipment, and you like class-related themes, it has no shortage of charm. His main stat is Wisdom, which gives him bonuses to Perception, and that's unfortunately almost all we can expect from him out of combat by default. However, you should expect that some builds will include a Monk level, to obtain some of its bonuses, such as Wisdom-based armor.

+ **Unique gameplay**

+ **Very fast and mobile**

+ **Swarms of blows**

+ **Can control enemies in melee**

+ **Becomes very robust at high levels**

- **Slightly weak damage, especially against targets with damage reductions**

- **Doesn't add much outside of combat**

- **Quite fragile at first**

❖ *So, which group composition to choose?*

There are hundreds of viable group compositions, we're not going to list them all. But if you follow the guidelines below, you should be able to overcome the majority of situations with some preparation and the right tactics:

● A sturdy melee fighter capable of dealing good damage (Barbarian, Paladin, Warrior, Cleric, Ranger, Monk)

● An offensive spellcaster (Wizard, Occultist, Sorcerer, Druid and Cleric to a lesser extent)

● A character capable of assuming a role similar to the Dodger, such as stealth, evasion, defusing traps (Rogue, Ranger, Bard)

● The face of the group, who will manage the dialogues, and who needs high Charisma if possible (Bard, Paladin, Occultist, Sorcerer)

● A healer/support, capable of strengthening the group, putting dying characters back on their feet and dispelling afflictions (Cleric, Druid, Bard, Paladin to a lesser extent)

The same character can play several roles. There are quite a few classes and especially hybrid subclasses. The Bard in particular can do almost anything. The important thing is to find the right classes to balance.

A good example of a group composed solely of original characters, which respects all the instructions above would be: Lae'Zel (Warrior), Shadowheart (Cleric), Astarion (Rogue) and Wyll (Occultist). Gayle can replace Wyll, but we lose the charismatic "face" of the group. Karlach can replace Lae'Zel.

Best Multiclasses

In old Baldur's Gate, it was possible to multiclass, and several notable characters used this feature, such as Jaheira and Imoen (Hey, it's me!). But there were significant constraints. Baldur's Gate 3 offers a much more free and intuitive system, allowing you to create really tempting builds, but there are still quite a few subtleties that will need to be taken into account for optimal results.

❖ Multiclassing in Explorer mode

Before we get into the details, you should know that if you play on the "Easy" difficulty mode aka Explorer, multiclassing is disabled. It must be said that this is not a simple feature, and it is possible to sabotage your characters by using it incorrectly. The developers had to disable it, figuring that players using this difficulty mode don't necessarily know what they're doing, and they don't really need it. If you want to multiclass in Explorer mode, simply change the difficulty in the menu. Nothing prevents you from returning to Explorer mode afterwards.

❖ How to multiclass?

It's very simple, from level 2, by clicking on the arrow under your character's portrait, the level gain screen appears. Instead of simply accepting the default result displayed on the screen, which suggests adding a level to your base class, click on the icon at the top right to add a new class that you can choose freely. The important prerequisite present in Dungeons & Dragons 5, which required having 13 in one of the characteristics linked to the additional class, has disappeared in Baldur's Gate 3.

Another huge advantage in Baldur's Gate 3 is that it's incredibly quick and easy to respec your character. If you make a mistake in your multiclassing, or want to try something else, simply pay 100 gp to Wither, or load the game. So there is no risk. Whereas on the table, if you try this, your game master risks laughing in your face, in the best case scenario.

❖ Advantages and limitations of multiclassing

Choosing an additional class allows you to automatically obtain the majority of its masteries. For example, by choosing Warrior, your Bard will learn how to use medium armor and the shield. But be careful, this is not as generous as if you had directly created a Warrior. In this case, there is no handling of heavy armor. You

152

must therefore pay close attention to the list on the right.

On the other hand, you do unlock the corresponding subclass by reaching the required level in your new class(es). As well as all the bonuses unlocked with the levels. The point of this is obviously to combine bonuses from different classes, for a devastating result. For example, many people will multiclass into Champion Warrior to get the critical strike chance, Spirit, and bonus attack at level 5, among other things.

We'll talk about spells later, but in the meantime, the main limit of multiclassing is obviously the maximum level, which is set at 12. The accumulation of all your classes cannot exceed this figure, which will necessarily limit your options. The proficiency bonus is based on your overall level, so this isn't a problem. But the Gifts are based on the levels of each class: you obtain a Gift every 4 levels of the same class. If you level up 12 levels of a single class, you receive 3, but if you level up 1 level of 12 classes, you receive none! Concretely, this tends to reward the fact of taking 4 or 8 levels of the same class, rather than getting lots of starting bonuses.

In summary, it is a delicate exercise, which requires you to find the right compromise between the starting masteries, the feats every 4 levels, and the special skills unlocked for each class and subclasses at certain specific levels, while seeking to obtain the best possible synergies.

❖ Additional attacks

Another important element when multiclassing is level 5 for the martial classes (Warrior, Barbarian, Paladin, Ranger, Barbarian and Monk), which gives an additional attack. It is not unlocked by combining several different martial classes. It's meta in a lot of builds, and it's also a huge power gain at low levels. It is strongly recommended not to multiclass before level 5 when using a martial class, so as not to delay obtaining it.

The Valor/Sword Bard receives it at level 6. The Warrior receives one more attack at level 11.

❖ Multiclass a spellcaster

The system is less favorable to spellcasters, and it becomes particularly technical. But in summary, you gain as many spell slots as with a single class, as long as you multiclass your character with pure spellcasting classes (Wizard, Wizard, Occultist, Druid, Bard, Cleric), while still getting a much greater variety of spells: the spell books are merged. Above all, you lose the possibility of reaching higher level spells. It takes 11 level of the same caster class to unlock level 6 spells, which only leaves one level available to multiclass in this case. As with martial classes, it is best to avoid mutating too early into a spellcaster, since level 5 corresponds to level 3 spells, such as Fireball.

● Some classes are considered semi-casters: Ranger and Paladin, it takes two levels for this to count as a spellcaster class.

- Finally, there are the thirds of spellcasters, these are the subclasses like Arcane Trickster and Eldritch Knight. It takes 3 levels for this to count as a caster class

Let us mention in passing that the spells of each class always use its main characteristic. So merging a Magician and a Sorcerer is not as profitable as one might think, since some spells will use Intelligence, and others Charisma. Getting 20+ in both of these traits without being made of wet toilet paper is difficult. It is better to favor classes that share the same characteristics:

- Charisma: Paladin, Sorcerer, Occultist, Bard

- Wisdom: Cleric, Druid, Ranger

- Intelligence: Magician...(Where is my Artificer?)

❖ Best Multiclasses

Needless to say, with 12 classes, 42 subclasses, and 12 levels, the possibilities are almost endless. And even good mixes are numerous. This becomes even more complex if we add the races and sub-races, since they bring their share of masteries and powers. There is no point in multiclassing into a Human Magician, if you want to unlock the use of light armor and shield, you already have them. There is therefore no universal recipe. The best thing to do is to look at a Wiki like https://baldursgate3.wiki.fextralife.com/Classes with the list of all the bonuses at each level, and choose what you want. Sometimes multiclassing can also be role play.

Here is a small list of very popular classes, of which you can take a few levels in order to obtain a big power bonus:

Warrior (1+) : The game's base class can be mixed with anything successfully, even Wizard, who will greatly gain survival with just one level. It is versatile, well-equipped and powerful. She even offers more Gifts than the others. Unless you're playing a spellcaster, it's pretty good.

- Level 1: Wearing medium armor, shield, Marian weapons, a fighting style, Second Wind

- Level 2: Fierce (additional attacks once per short rest)

- Level 3: Subclass

- Level 4: Gift

- Level 5: Additional attack

- Level 6: Bonus Gift

- Level 8: Gift

- Level 11: Another additional attack, are you sure you want to multiclass?

- Level 12: Gift

Barbarian (1+) : This class offers some unique bonuses that the Warrior lacks, such as Rage for martial classes and Reckless which gives advantage on attacks. Investing 3 to 4 levels in Barbarian is therefore a good plan. There is also Unarmored Combat, which a lot of classes like spellcasters may want to gain survival with a single level, since Constitution becomes armor.

Dodger (3+): Far from being a simple locksmith, the Dodger offers both huge damage gains with the sneak attack, tactical tools like stealth and disengagement in combat, utility skills, and even survival. The Assassin in particular is popular, whether in melee, or as an Archer, combining it with Ranger or Warrior levels for the extra attack. The main thing to remember is that the damage from the sneak attack increases every odd level of Dodger, so investing 5 to 7 levels in it is a good plan.

Cleric of the Lore of Life or Lore of War (1+) : The first level of the Cleric is incredibly generous, especially choosing these two areas which allow you to wear heavy armor and martial weapons in the second case. A second level is also very profitable, if your character has a lot of wisdom. The area of   life will make your character an excellent healer, who will also prove to be very robust. This can be a good choice for a Druid or Paladin.

Paladin (2+) : Perhaps one of the classes entitled to the most bonuses during its first levels. Like the warrior he wears armor, shields and weapons. What makes the Paladin so attractive is the ability to perform Divine Smites by sacrificing his spells starting at level 2. This makes him a huge damage boost for all melee hybrid classes, like the Bard, Ranger, 'Occultist, Cleric, etc. The fighting style at this level cannot be refused either.

Multiclass and specializations

A class is more than just a profession in Baldur's Gate 3, it is a way of living and acting, as the origins, skills and advantages of each class tend to be different. Adventurers have the choice between numerous classes which allow them to approach combat, and the various situations which can be encountered in many ways. The classes that make up your party will often determine what options are available or not, and how to approach fights. A warrior will rely on his physical abilities and intimidation to solve his problems, while a rogue will perhaps try to turn a situation to his advantage through deception, or he may simply relieve his target of the object longed for.

❖ List of classes

All D&D5 base classes and their specializations will be available in Baldur's Gate 3 upon release, along with 46 subclasses. They more or less resemble the rules of the game on paper. However, you should know that various adjustments have been made by Larian Studios, in particular for the most "role play" skills which are very difficult to transpose into a video game.

It would take several pages to detail each class. If I can't tell you everything here, here is the list of classes that are present with a short description of their main faculties:

- **Warrior**: The classic but formidable basic fighter, because he is robust as well as the absolute master of weapons and armor. He can easily turn into a killing machine.

- **Dodger**: The famous thief, able to camouflage himself, empty pockets, pick locks, spot traps and perform a devastating sneak attack in melee or from a distance.

- **Cleric**: An armored cleric who casts divine spells like heals and buffs, or more offensive spells as needed.

- **Druid**: A defender of balance who casts natural spells (healing, vines, animals and who transforms into a wild animal.

- **Magician**: The nerd who needs to prepare his arcane spells to cast them, but he has an immense repertoire and great knowledge.

- **Paladin** : A holy fighter who combines martial mastery with divine techniques and spells.

- **Lurker**: An adept at survival in the natural environment, competent in melee, but above all he is a potential virtuoso with the bow.

- **Monk**: An agile and fast fighter who can fight with his bare hands and without armor.

- **Barbaric**: A furious and reckless fighter, often without armor, but who can inflict immense physical damage and take quite a bit of it.

- **Took place**: A jack of all trades who inspires the group, casts a few spells, and uses some skills similar to the Dodger.

- **Sorcerer**: A highly adaptable arcane spellcaster who can cast and modify his spells without having to prepare them, but his repertoire is greatly reduced.

- **Occultist**: A "new" class that has made a pact with an external, non-divine entity (such as a demon) to obtain its arcane spells and powers.

Not all classes are represented by your companions, but you can pay Respec to Withered to reassign their characteristics and change their base class, even if it's not very role play.

❖ Class restrictions

If you've played the first Baldur's Gate and the Enhanced Editions in particular, the majority of the classes should be familiar to you, although there are some new ones There are some major changes from previous games, however. For starters, there aren't really any racial or alignment restrictions anymore. There are still recommendations, but they are no longer strict obligations. You are no longer forced to be a Lawful Good Human if you want to play a Paladin, this will certainly upset some, but on the other hand, it opens up many options in terms of gameplay, but also role play.

❖ Maximum level and progression

Many may cringe, but it has been announced that the maximum level in Baldur's Gate 3 will be level 12. For the record, Baldur's Gate 1 and its expansion allowed you to reach a level ranging from 7 to 10 at the time in depending on your class. In the context of D&D5, this corresponds to an intermediate level, the maximum being 20. This allows you to have significant development of your class and to unlock good spells and techniques, but without entering into high-level content. Note in passing that all classes need the same amount of experience to level up this time.

❖ Specialization of classes, aka subclasses

In addition to all the changes related to the D&D5 rules, be aware that it is generally no longer possible to directly choose a class kit at level 1, such as Specialist Wizard, Assassin Thief, Archer Ranger, Kensai Warrior, Paladin Rider for n 'to name just a few notable ones. You start with a basic class in the majority of cases, and a specialty/track/school is chosen at level 2 or 3 depending on the class. This is where your Magician can choose to be a Necromancer, your Druid to be a shapeshifter, etc. Each specialty introduces new talents, bonuses and mechanics throughout the levels, in addition to the basic ones of the class. Some classes still have a choice of origin to make for role play reasons.

There are still some notable exceptions with the Clerics who choose a deity and its domain at level 1, the Occultist who also chooses his boss at creation at level 1, and the same goes for the Sorcerer with the source of his magic.

There are a total of 46 subclasses in Baldur's Gate 3. There are at least 3 per class, and the Wizard has 8.

❖ List of archetypes and specializations

The term "specialization" is not necessarily the most appropriate, because the name of the choice to be made is different for each class. In any case, here is the basic list, some classes have more choices than others. The name of the specialization is also more or less telling, we will detail all this in articles dedicated to each class. Here is their list, with an extremely brief and simplified, even simplistic summary of their concept.

Note that in almost all cases, the classes present have 3, but it can be many more.

- **Warrior**: Warmaster (combat maneuvers & tactics) or Eldritch Knight (spells and link with his magic weapon) or Champion (bonus to critical and physical abilities). This 3rd option is absence of early access.

- **Dodger**: Thief (dexterity, theft, stealth) or Arcane Trickster (spells), or Assassin (assassination & infiltration). This 3rd option is absence of early access.

- **Cleric**: Choice of a domain at level 1 with associated spells & skills: Life,

Deception, Light, (Absent: War, Nature, Knowledge, Storm, Forge).

- **Druid**: Circle of the Moon (metamorphosis into a more fearsome animal) or Circle of the Earth (spells).

- **Magician**: Choice of an arcane tradition linked to a school of magic, with original bonuses specific to each. Only those of Abjuration and Evocation are present. The others are: Divination, Enchantment, Illusion, Necromancy, Transmutation. No longer blocks an opposing school.

- **Paladin** : Oath of Devotion (conventional sacred effects), Oath of the Elders (natural and elemental effects), Oath of Vengeance (combat bonus against an enemy). Not forgetting the perjured paladin.

- **Lurker**: Hunter (damage and attack bonus in melee or ranged) or Beastmaster (companion animal in combat). But the class looks like it got some major overhauls in BG3.

- **Monk**: Way of the Open Hand (unarmed combat), Way of the Shadow (stealth and control), Way of the 4 Elements (elemental spells).

- **Barbaric**: Path of the berserker (bonuses linked to rage & bonus attacks) or Path of the totem warrior (bonus linked to the chosen totemic animal).

- **Took place**: College of Knowledge (skills & spells) or College of Valor (Inspiration, combat bonus & spells in armor).

- **Sorcerer**- Choice at level 1: Draconic Ancestor (resistance & elemental link) or Wild Magic (reroll rolls and random effects).

- **Warlock**- Choice of an overworld boss at level 1: Fiend (infernal spells), or Great Elder (control). The Archfey (fairy spells) is absent.

❖ *Multiclassage*

This very popular feature among fans of optimization and experimentation will be present in BG3, it allows you to gain levels in several different classes and therefore combine their abilities. The result can sometimes be very powerful, or very interesting at least in terms of role play. This is a feature often used by players who want to play the game solo (in early BG at least), or who simply want to expand their range of options. As the groups will be much smaller in Baldur's Gate 3 (only 4, compared to 6 for the old ones), it can be difficult to integrate everything you want. The Dodger class is a popular choice for multiclassing, since it allows you to get its incredibly desirable lock picking, pickpocketing, and trap disarming abilities but without having to take a "pure" one.

Multiclassing is really simple and flexible in D&D5, and it's even more so in Baldur's Gate 3, since the ability requirements have been removed.

With each level you gain, you can freely choose to unlock a new class. You can create your Barbarian Magician if you want, but remember that the level 12 limit applies to your total classes. This can easily deprive you of powerful high-level abilities. This is not to be done lightly.

This affects all classes, including those that were previously unavailable for multiclassing in previous editions (like Paladin and Monk). What's more, it is even possible to choose an additional 3rd or even 4th class later. You can choose which class to assign your additional levels to later, you are free to have a big imbalance if you wish, such as 5 warrior and 1 rogue. Be careful though, it is the overall level that is taken into account to determine the experience required for the next level, the mastery bonus and the level limit. The cumulative level of all your classes cannot therefore exceed 13 in this case. As classes all gain a very useful special feat at level 10, multiclassing often results in depriving yourself of high level and raw power bonuses, in order to gain versatility.

❖ Respec/Reset Levels

Baldur's Gate 3 will offer you the opportunity to reset your levels, in order to correct your mistakes and experiment. A certain NPC at the camp, encountered during Act 1, will offer you this option: Withered. You will be able to choose your classes, gifts, specializations and spells again. This is true for both your protagonist and your companions.

Respecialization

It's not something a game master would allow you to do on the tabletop, but Baldur's Gate 3 allows almost unlimited resp for everyone. But you have to advance in the story, and unlock a certain NPC before that becomes an option.

❖ How to unlock character respecs?

You must complete the introduction aboard the Nautiloid, and begin the first act of the game. You must enter the beach crypt to access the required NPC. There are several ways to enter:

● Unlock the door on the beach, which requires thief tools.

Otherwise, you have to take the grand tour, and reach the top of the cliff with a group of looters and several alternative ways to enter:

● Go through the trapdoor at the end of the path, on the side of the cliff.

● Drop the hanging rock onto the mosaic.

● Convince the doorman to let you in.

❖ Find the secret room of the Damp Crypt

● Once in the crypt, reach the main room with inanimate skeletons everywhere, and a large statue of Jergal.

● Save your game. Then, move to the back left of the room, at the end of the dark corridor. Your characters will make a perception check. If everyone misses it, load your game.

● If successful, a button will appear on the wall. By clicking on it, the skeletons in the room will wake up and attack you, which will trigger a fight. Disarming them beforehand and positioning your characters in advance can help.

● After the fight, open the sarcophagus, which will release Blight, a fairly mysterious undead, but who is not hostile. It doesn't matter too much what answers you give him, as long as you don't attack him.

❖ Change the class of companions or your original character

● After speaking to Wither, go to the camp by clicking on the icon at the bottom right of the interface. Withered will be present in the area.

● Go talk to him with the character whose class you want to change (or reset the levels). You just have to group the corresponding character, then take direct control of it.

● Now all you have to do is ask him to help you change classes, which costs 100 gold each time.

This option has the major advantage of being able to change the class of the original Characters, and even of your companions. This allows you to bring the characters you prefer into combat with you, without harming the synergies of the group.

But be careful, the personal history of each character is partly linked to their class, like Halsim the bear druid. This may prove annoying during certain passages.

Best donations

With Baldur's Gate 3, Larian Studios made the daring choice to impose real role play limits on the Paladin class, unlike the old titles in the license. It is no longer enough to maintain a high level of reputation, or a certain alignment to maintain this very powerful and prized class. Indeed, depending on your choice of oath, when creating your Paladin, you are forced to follow certain guidelines in your actions. The abuse that the majority of RPG players naturally indulge in, "murder hobo-ism", aka killing and looting everything without getting caught, is not an option, and you really have to be careful about your actions and their consequences repercussions. The good news is that if you want to play a black knight rather than a white knight, you will quickly have the opportunity to change your vocation.

❖ How to break your oath quickly?

One of the first opportunities to begin your descent towards the dark side is available at the end of the tutorial, after waking up on the beach. As you advance in the area, you will come across two Tieflings who have imprisoned Lae'Zel. If you don't conduct the negotiations well, or if they fail, you will have to make a painful choice between killing these two innocent idiots, or your potential mate. Killing Tieflings is considered a breach of your oath, which will have repercussions. It must also be said that Lae'Zel is racist, cruel and aggressive at the same time, he is not really a good person to begin with.

❖ Perjured Knight

Then return to the camp, by clicking on the Campfire icon, at the bottom right of the interface. A new NPC will be waiting for you there, the Forsworn Knight, a kind of dark armor. He will explain the situation to you after the betrayal of your oath.

The discussion can be oriented in different ways, and it is even possible to make amends to return to your original oath. But if you persist in this path, you can choose to break your oath completely and become a Forsworn Paladin.

A small cutscene follows, which shows that you are really joining the dark side, like this brave Anakin. The new subclass is then unlocked. Rather than Radiant damage you will inflict Necrotic damage in addition to giving the advantage against the affected target for example. By gaining levels you will unlock new abilities, such as summoning the undead.

Forsworn Paladin

With Baldur's Gate 3, Larian Studios made the daring choice to impose real role play limits on the Paladin class, unlike the old titles in the license. It is no longer enough to maintain a high level of reputation, or a certain alignment to maintain this very powerful and prized class. Indeed, depending on your choice of oath, when creating your Paladin, you are forced to follow certain guidelines in your actions. The abuse that the majority of RPG players naturally indulge in, "murder hobo-ism", aka killing and looting everything without getting caught, is not an option, and you really have to be careful about your actions and their consequences repercussions. The good news is that if you want to play a black knight rather than a white knight, you will quickly have the opportunity to change your vocation.

❖ How to break your oath quickly?

One of the first opportunities to begin your descent towards the dark side is available at the end of the tutorial, after waking up on the beach. As you advance in the area, you will come across two Tieflings who have imprisoned Lae'Zel. If you don't conduct the negotiations well, or if they fail, you will have to make a painful choice between killing these two innocent idiots, or your potential mate. Killing Tieflings is considered a breach of your oath, which will have repercussions. It must also be said that Lae'Zel is racist, cruel and aggressive at the same time, he is not really a good person to begin with.

❖ Perjured Knight

Then return to the camp, by clicking on the Campfire icon, at the bottom right of the interface. A new NPC will be waiting for you there, the Forsworn Knight, a kind of dark armor. He will explain the situation to you after the betrayal of your oath.

The discussion can be oriented in different ways, and it is even possible to make amends to return to your original oath. But if you persist in this path, you can choose to break your oath completely and become a Forsworn Paladin.

A small cutscene follows, which shows that you are really joining the dark side, like this brave Anakin. The new subclass is then unlocked. Rather than Radiant damage you will inflict Necrotic damage in addition to giving the advantage against the affected target for example. By gaining levels you will unlock new abilities, such as summoning the undead.

BUILDS DE CLASSES

Build Sorcerer

For players new to playing Dungeons & Dragons 5th Edition, it can be difficult to guess the best way to optimize each of the 12 classes. It's even more difficult if we take into account the 46 possible subclasses. Here we offer you the main guidelines to follow and the best choices to make in general in Baldur's Gate 3. Nothing prevents you from modifying this build according to the race you wish to play, and the objects available.

It is possible to respecialize your character from a certain point in the adventure, from the camp. Don't be afraid to make mistakes.

❖ *Presentation of the Sorcerer*

Those who wish to play a spellcaster, without the complexity of the Mage, nor the role-play of an old bearded man in his tower, can turn to the Sorcerer. A class capable of unleashing more spells than the others, but also of modifying their effect with its powers, in order to gain flexibility. His spell repertoire is much smaller than those of the Wizard, but the elements listed above make up for it. He also has access to very different specializations, such as Draconic Bloodline, Wild Magic and above all, Storm Sorcery.

Another advantage of the Sorcerer is that he uses Charisma to cast his spells, making him a particularly attractive choice for the protagonist.

+ **Lots of spells**

+ **Metamagics**

+ **High charisma for dialogues and diplomacy**

- **Quite fragile**

- **Clearly does not replace a Magician in terms of skills**

❖ *Sous-classes*

The 3 subclasses of the Sorcerer have very notable particularities. It is also one of the rare classes that can directly choose its subclass at level 1.

● **Wild Magic** : Grants advantages and power bonuses, at the cost of sometimes unpredictable effects on your spells.

● **Dragonblood** : Draws its power from the blood of dragons. This gives him various survival bonuses, and spell abilities related to dragons.

● **Storm Sorcery**: Use the power of lightning and wind to kill your enemies

162

while flying on the battlefield.

❖ Origin

The original character "Dark Urges" is a Storm Sorcery Sorcerer by default (although it is possible to change it). This makes it a pretty good choice if you plan to play this class. There is no companion with this class by default.

❖ Historical

It is better to choose a History conferring skills linked to Charisma, such as Guild Artisan (Persuasion and Investigation), Charlatan (Deception, Sleight of hand), or Sage (Arcana, History) to compensate for the possible absence of a Magician in the group .

❖ Race

As all races now benefit from the same characteristic bonuses (+2 and +1 to be allocated to the choice), the choice of a race is much less important than before. This can nevertheless have an impact, by granting useful bonuses and resistances.

● **Human** : If you are not playing the dragonline subclass, this allows you to equip light armor and a shield. The extra skill is also good to have.

● **Dragons** : What's more normal for a wizard than to have double dragon blood This race will grant you an elemental resistance of your choice and a breath, which will complete your arsenal. Warning: Do not choose a blue or bronze Drakeid if you plan to play Storm Sorcery, that would be redundant.

● **Tieffelin** : Fire Resistance, Darkvision, and the additional cantrip are all interesting.

● **Nain d'or or Duergar**: +1 Life Point per level, Darkvision and Poison Resistance are all good to take. The other dwarf subraces are also interesting in their own way, particularly the Duergar with improved Darkvision and bonuses to saving throws.

● **Skin elf** : The added perception mastery, charm and sleep protection, and cantrip make this a viable choice, especially if you like having a "handsome" character.

❖ SKILLS

The priority is to obtain mastery of at least one or two skills related to Charisma, with Persuasion first, then Deception and Intimidation. You can then look to gain other useful skills like Perception.

❖ Features

Like many similar classes, the Sorcerer relies on Charisma, Dexterity and Constitution, in that order of priority. Your points distribution should look something like this:

● Force : 8

- Dexterity: 14 to 16
- Constitution: 14 to 16
- Intelligence : 8
- Wisdom: 8 to 10
- Charisma: 16 to 17

❖ Equipment

As a spellcaster, the Sorcerer doesn't worry too much about equipment until he gets his hands on magical items that provide useful bonuses. If you have a good Dexterity score, you can use a Light (or Heavy) Crossbow at the start, it will hurt more than your cantrips.

- You will not wear any armor at all, or just light armor, if your race unlocks this mastery (which we recommend).
- Use a staff, or any weapon and a shield, if you have the required proficiency. Extra armor is always welcome when you're this fragile.

❖ Sorts

Spell choices may vary depending on personal preference, party composition, and subclass. This is especially true if you are playing with Storm Sorcery, since you will be strongly encouraged to sacrifice one or two spells of each level in order to trigger the special effects linked to it.

- **Cantrips** : Fire Bolt, Frost Ray, Icy Cold, Light, you can also try to fit Acid Spray or Electric Grip if you have additional spells.
- **Level 1** : Magic Missile, Thundering Wave, Color Orb (Storm Sorcery)
- **Level 2** : Shatter, Burning Ray
- **Level 3**: Fireball, Haste, Lightning (Storm Sorcery)

❖ Gameplay

This will greatly depend on your level and the spells chosen, but as a general rule, you will rather hide at the back of the group to cast your spells from afar, while your more robust companions occupy the enemy.

If you favored shorter range spells and/or you are playing Storm Sorcery, then the gameplay changes a bit. You will be more likely to stay close to enemies in order to position yourself in a way to cast your area spells in the most effective way possible, before flying away.

❖ Leveling et dons

Baldur's Gate 3 does not ask you to make very complex decisions during leveling, but some important choices still await you, such as metamagic at level 3 (see below).

Every 4 levels, you must choose between 2 characteristic points to attribute, or a gift of your choice. The option will appear 3 times in total.

- Aiming for 18 or even 20 Charisma is strongly recommended.

- **Elemental Adept** : Potentially interesting if you play Storm Sorcery. By choosing electricity or thunder, you increase your damage a little, and eliminate enemy resistance.

- **War Mage**: Avoid losing concentration in the event of damage received. Also allows you to perform attacks of opportunity with Electric Grip. This is another attractive option for Storm Sorcery, which tends to be closer to melee.

- **Metamagic Expertise** : Allows you to choose an additional metamagic and gain 2 additional sorcery points. If you want to expand your options.

- **Magic Artillery** : If you prefer to cast your spells from a very long distance, well sheltered, this is the gift you need. This will double its range and reduce the penalties on the chances of hitting.

❖ Metamagic

At level 3 and 10 you can choose a metamagic:

- **Twin Spells** : Allows you to send a single-target spell on two targets. This makes it by far the meta option.

- **Transmutation (Storm Sorcery)** : Allows you to change the element of the spell, and therefore trigger the subclass bonus.

Build Magician

For players new to playing Dungeons & Dragons 5th Edition, it can be difficult to guess the best way to optimize each of the 12 classes. It's even more difficult if we take into account the 46 possible subclasses. Here we offer you the main guidelines to follow and the best choices to make in general in Baldur's Gate 3. Nothing prevents you from modifying this build according to the race you wish to play, and the objects available.

It is possible to respecialize your character from a certain point in the adventure, from the camp. Don't be afraid to make mistakes.

❖ Presentation of the Magician

The ultimate fantasy class archetype is probably the wizard. We can't help but think of an old bearded man who casts spells like Merlin or Gandalf. The Magician is the intellectual of the group, he has great knowledge in many areas, as well as the largest repertoire of spells of all the classes. If he has the right spells in his spellbook, and he has time to prepare them the day before, he can deal with any situation. However, this has its limits, and the Magician's spells are not designed to heal, for example. They shine especially for controlling and burning enemies. The Magician is nevertheless a class with great weaknesses. He cannot wear armor by default and he has little health, making him very fragile. Additionally, he can only cast a limited number of spells per day. All of this makes him an absolutely fantastic class to have in a group, but he's going to need the protection of his teammates to really shine.

+ **Large spell repertoire**

+ **Big advantage over the enemy if you know what spells to prepare**

+ **Only class with Intelligence in the game and the skills that go with it**

+ **Excellent area damage**

+**Fantastic control spells**

- **Tragically fragile**

- **Minimal combat utility once his spells are exhausted**

- **Must choose your spells before resting**

❖ Subclass: Evocation

The Magician is the king of specializations, aka arcane traditions, since he has no less than 8, one per school of magic. This no longer works like it did in older editions of Dungeons and Dragons. Each specialization has its own advantages, but it does not grant additional spells, and it does not prevent you from casting spells from other schools. Arcane tradition is chosen at level 2.

Each Magician subclass has its charm, but it is generally advisable to choose Evocation. It is by far the most represented and most versatile spell school when you want to inflict damage. It will allow your Magician to do more harm, and above all, to do so without killing his companions. You can throw your Flaming Hands and Fireballs into the fray, without turning the idiot paladin into smoking ashes! No other spellcasting class has this privilege.

❖ Origin

One of the original characters in Baldur's Gate 3 is a human Magician: Gale. This makes it a good choice for learning about the class as well as its unique history. You can also create a custom character, in order to have more freedom in your choice of race and other elements.

❖ Historical

This will depend on your race and group. But if you plan to play solo, then your Magician will be the protagonist, which will encourage you to make some unusual choices, in order to have a better chance of persuading your companions to grant your requests:

● Recommended: Noble (Persuasion and History) Guild Artisan (Persuasion and Insight)

● Alternative: Sage (Arcana and History)

166

❖ Race

As all races now have the same characteristics (+2 and +1 to freely assign) you can choose your race based on additional bonuses, or its aesthetics, without suffering too much.

- **Githyanki**: Gives mastery over light and medium armor, which is a good survival gain. The mage hand cantrip is also useful. Finally, the ability to know Astral allows you to have mastery of all the skills of a characteristic, which is really fantastic for the protagonist. This will help you greatly on your Persuasion rolls and such. In return, you will look like a toad.

- **Human** : Grants proficiency with light armor and shield, and an additional skill.

- **Half High Elf** : Darkvision, resistance to charm and sleep, as well as an additional Minor Spell of your choice. You also get to wear light armor and shield as for humans.

- **Skin elf** : The classic choice, with a minor spell, resistance to Charm and sleep darkvision, and finally, mastery of Perception, which is always good to take.

- **Nain d'or** : If the miserable survival of the Mage horrifies you, make a dwarf. Poison Resistance, and the extra hit point per level should help. Night vision is also good to have.

- **Tieffelin** : Here, it's Fire Resistance, a cantrip and Darkvision is also an excellent choice. Fire damage is probably the most common.

❖ SKILLS

This will depend on your History. If you took the Guild Artisan or Noble, you have access to Persuasion, which is something you will really need as a protagonist.

You can take the rest depending on what the group will master, but as a Magician, here are your priorities:

- Arcana (automatic)
- History
- Investigation
- Religion

❖ Features

The 3 important characteristics of the Magician are Intelligence, Dexterity (for survival), and Constitution (ditto). But if you don't like spending your time loading a save after a failed Persuasion roll during a dialogue with a companion, we advise you to also invest in Charisma. That makes a lot. Aim for the following scores:

- Force : 8
- Dexterity: 14 to 16

- Constitution: 14 to 16
- Intelligence: 16 to 17
- Wisdom: 10
- Charisma: 12 to 14

❖ Equipment

There's not much to say here, the Wizard no longer normally uses his weapon to attack since the addition of cantrips. You will normally equip the first staff offering bonuses to the spells you find. The same goes for the mage robe. If you ever play a race that offers armor usage bonuses, then choose the heaviest armor possible, while preserving the dexterity bonus.

Early in the game, you can use a crossbow instead of your cantrips, if your dexterity is good.

Lump the Enlightened, the Ogre mage of the Dilapidated Village drops a Crown offering 17 Intelligence, during chapter 1. Some players count on it to have an artificially high Intelligence score, which allows them to neglect this characteristic for invest their points elsewhere.

❖ Sorts

You'll learn many more spells later, but here are the most useful ones to choose immediately when they become available:

- **Cantrips**: Fire Bolt (versatile and can ignite or explode barrels and other objects), Mage's Hand (activates traps and others), Light, Icy Contact (alternative spell at a distance, little resisted and formidable against the undead)

- **Level 1**: Magic Projectiles, Grease, Sleep, Tasha's Laugh, Thundering Wave. Afterwards, take Burning Hands, Light as a Feather and Chromatic Orb.

- **Level 2** : Burning Ray, Foggy Step, Shatter, Swarm of Daggers, Immobilize People, Web

- **Level 3** : Fireball, Animate the Dead, Flight, Hypnotic Pattern, Counterspell

❖ Gameplay

This will depend greatly on the spells prepared, but generally, the Magician's objective is to stay out of danger by placing himself behind his companions, or in a position that is difficult to access. And cast spells to control or kill enemies, depending on the circumstances.

It has the best terrain control spells in the game at each level, like Grease, Web, and then Hypnotic Pattern, which can theoretically slow down a whole group of enemies, while the rest of the group lets them come, and kills one at a time. one those who managed to pass. He can also combine them with persistent damage spells (Cloud of Daggers) or a direct AoE like Fireball. The Magician can then easily cast spells to finish off several weakened enemies at once with Magic Missile and Burning Ray for example.

❖ Leveling et Dons

Every 4 levels, you will have the choice between +2 Characteristics to freely assign (Maximum 20 points in a characteristic) or a Gift. You will therefore be entitled to 3 Gifts or +6 Characteristics at most (or a mixture of the two).

Raising your Intelligence up to 20 is tempting and profitable, but there are also some very useful Gifts:

- **War Mage**: Avoids losing concentration, and gives the possibility of casting spells instead of attacks of opportunity.

- **Offensive Mage**: Doubles the range of attack roll spells, allows you to ignore cover and grants an additional cantrip. This transforms the Magician into a sniper hiding at the other end of the area.

- **Elemental Adept** : Ignores resistances to an element type of your spells, and slightly increases its damage on average. An often recommended choice if you like to throw fireballs and fiery rays, since Fire is the most resisted element in general.

Build Occultist

For players new to playing Dungeons & Dragons 5th Edition, it can be difficult to guess the best way to optimize each of the 12 classes. It's even more difficult if we take into account the 46 possible subclasses. Here we offer you the main guidelines to follow and the best choices to make in general in Baldur's Gate 3. Nothing prevents you from modifying this build according to the race you wish to play, and the objects available.

It is possible to respecialize your character from a certain point in the adventure, from the camp. Don't be afraid to make mistakes.

❖ Presentation of the Occultist

Veterans of Baldur's Gate and Dungeons & Dragons games will certainly be familiar with the Sorcerer/Wizard, but not necessarily the Occultist, who is a more recent addition. He is a spellcaster also based on Charisma, but he draws his powers from a pact made with a superior entity, such as a powerful fiend, an archfey or a great elder. This gives him abilities quite different from the Sorcerer and the Mage, with several possible orientations, in exchange for a much lower spell pool. But the most popular is undoubtedly the Occult Explosion spec, which consists of continuously casting a minor spell, but overpowering for its level. Even if you fight one after another, you will remain effective.

+ An excellent spell available in unlimited quantities

+ Relatively sturdy for a spellcaster

+ Based on Charisma, therefore perfect for the protagonist

- Very limited reserve of normal spells

❖ Subclass: Overworld Boss

The Occultist is one of the rare classes directly choosing its subclass during its creation. This is rather logical, since it determines with what type of entity he signs his pact. The powers granted change depending on this choice.

- **Fiend** : Survival bonus based on temporary hit points. Also gives more offensive spells, and around fire afterwards like Fireball.

- **Great Elder** : Chance to scare enemies on a critical hit. Grants control spells like Tasha's Laughter.

- **Archive** : Can scare or charm enemies with an action. Grants illusion and chaos spells like Faerie Glow.

❖ Origin

One of the original characters is an Occultist: Wyll. This is a good choice for your Occultist, since his story centers around his interactions with his otherworldly boss a ruthless she-devil. After all, obtaining supernatural powers is not free, and they must be earned. This allows you to experience the role-playing aspect of the class, whereas creating a custom character risks depriving you of that.

❖ Historical

Wyll's history is frozen. But if you create a custom character, you are free to choose Try taking an option that gives you useful skills, like Guild Artisan (Persuasion, Investigation), or Charlatan (Deception, Sleight of Hand).

❖ Race

Larian has rebalanced the races in the game, they all have +2 and +1 in two characteristics of your choice. So you are not forced to play Half-Elf or Tiefling. Each race nevertheless has other bonuses, a little more secondary, but still very useful. Here are some breeds that stand out:

- **Human**: One more skill to choose from, and grants shield handling.

- **Tieffelin** : Fire resistance and Darkvision

- **Dragons**: A mini-dragon breath and resistance to the element of your choice

- **Nain d'or** : Dark vision, resistance to poison, and one additional hit point per level, which provides excellent survival, at the cost of reduced movement distance.

❖ SKILLS

As Charisma is by far the highest trait in the class, and you will normally be playing this character as the protagonist, it is best to invest in the associated skills:

- Persuasion
- Deception

- Intimidation

You can optionally choose other skills depending on the composition of the group and the points available:

- Perception
- Arcana

❖ Features

There are only 3 important characteristics for the Occultist: Charisma above all, then Dexterity and Constitution in order to improve his survival. The breakdown should look something like this, depending on which characteristics you are willing to sacrifice:

- Strength: 8 to 10
- Dexterity: 14 to 16
- Constitution: 14 to 16
- Wisdom: 10
- Intelligence : 8
- Charisma: 16 to 17

❖ Equipment

As a spellcaster, this doesn't matter too much. Equip the best light (or intermediate in some cases) armor allowing you to benefit from your full Dexterity bonus.

Unless you're playing a "blade" build, which encourages using a rapier, equip the weapon of your choice and a shield if you're proficient with it. A staff is also an option if you find a magical one with good bonuses. Baldur's Gate 3 doesn't worry about the issues with having your hands free to cast spells, as long as you have proficiency with the equipment.

❖ Sorts

Minor spells: Occult explosion (absolutely vital) and a spell of your choice, Friend can be interesting if you like to manipulate your interlocutors, regardless of the consequences. Mage's Hand and Weapon Protection are useful if you use them well

- **Level 1**: Algathys Armor and Hex
- **Level 2** : Not foggy, Shattering,

❖ Gameplay

The gameplay is quite very simple. Cast Algathys Armor at the start of the day. At the start of the fight, use your control spells, or your most powerful offensive spells on the most dangerous enemies, then send Eldritch Discharge each turn. The spell's range and damage make it viable in almost any situation. Use Hex as quickly as possible on bosses to increase damage.

❖ Leveling et dons

Level 2 allows you to choose 2 occult manifestations, others arrive at level 5, 7 and 9 and 12. For this build, those that we advise you to take as a priority are:

- **Heartbreaking discharge** : Adds the charisma bonus to Eldritch Blast damage, which is absolutely fundamental.
- **Repulsive discharge** : Repels creatures hit by Eldritch Discharge, allowing enemies to be kept at bay or caused to fall into the void.

At level 3, you will be able to choose a pact: Blade (makes your magic weapon), grimoire (a few more spells in your repertoire) or chain (summon a familiar). The familiar will generally be the most useful choice.

You will be able to choose a gift every 4 levels, or +2 characteristic points to freely assign. You will have 3 opportunities to do this (level 4, 8 and 12), so you have to choose carefully.

- Aiming for 20 base Charisma is obviously a good option.
- **Magic Artillery** : Gives an additional cantrip, and increases the critical strike chance of spells that can be (like Eldritch Explosion).
- **War Mage** : Reduces the chance of losing concentration, while adding a spell cast as a melee reaction.
- **Elemental Adept** : By choosing the element of your favorite spells (probably fire), you will optimize their performance in all situations. But it's much less interesting

Sword Bard

For players new to playing Dungeons & Dragons 5th Edition, it can be difficult to guess the best way to optimize each of the 12 classes. It's even more difficult if we take into account the 46 possible subclasses. Here we offer you the main guidelines to follow and the best choices to make in general in Baldur's Gate 3. Nothing prevents you from modifying this build according to the race you wish to play, and the objects available.

It is possible to respecialize your character from a certain point in the adventure, from the camp. Don't be afraid to make mistakes.

❖ Introducing the Sword Bard

In older editions of Dungeons & Dragons, the Bard wasn't very popular, and that was also the case in Baldur's Gate. It must be said that the class was really difficult to use correctly. Baldur's Gate 2 nevertheless introduced a viable option and an interesting companion in the form of Haer'Dalis, a Sword Bard. Baldur's Gate 3 uses the 5th edition rules, and the Bard has changed a lot, it is now a very interesting class, thanks to its unparalleled versatility. She is capable of fighting well and casting spells, although she is obviously not the most effective in each of these areas. It is above all the fact that she also masters an immense amount of skills, and that she is also able to strengthen her companions with her songs, which makes her attractive. The Bard is therefore the ultimate jack of all trades, with a few unique specialties to boot. All this allows him to replace quite a few classes in the group, such as the Thief or the Magician, at least in terms of skills.

The Sword Bard is the subclass (or bardic college) that specializes in melee combat using two weapons or a shield, which greatly enhances its role as a physical DPS. You can also take the Value subclass, to strengthen the Support aspect and tanking.

+ Incredibly versatile

+ Can potentially become one of the most powerful builds

+ A great choice for the protagonist

- Clearly not the best martial class

- Need to multiclass Paladin to really shine in combat

❖ Sous-classes

The Bard has three subclasses unlocked at level 3, which will direct him towards slightly different roles, depending on your choice. This gives you time to decide which branch to choose:

● **Know**: Greatly strengthens the skills of this type, and further predisposes the Bard to a role as a spellcaster, by multiclassing him Occultist for example.

● **Value**: Strengthens both the Bard's survival in melee with a shield, and his ability to strengthen his companions. Many players consider this to be a superior subclass to Sword Bard, especially if you don't plan on multiclassing.

● **Sword**: You gain attack, mobility and defensive abilities in melee at the expense of the Support aspect. Even with all this, she is not equal to a martial class.

❖ Multiclassage

This is not an option that will please everyone, especially during their first game. But if your goal is to have as formidable a Bard as possible, then this is a mandatory step.

The recommended multiclassing is Paladin Oath of Vengeance, and take at least 2 levels in this class. You can then take 10 levels of Sword Bard, which will still grant you the unique bonuses of the class.

The reasons behind this multiclassing are also multiple: The Paladin is entitled to an outrageous number of bonuses and skills during these levels, such as the use of heavy armor, the shield, martial weapons, etc. Best of all, he gains the ability to convert his spell slots into divine smites with his melee attacks, which will allow your Sword Bard to perform plenty of them in combat. It can be interesting to go up to 4 Paladin levels, to unlock its additional bonuses.

❖ Origin

There are no Bards in the Origin classes, and the races available in the Origin classes are not ideal for a Bard. You can choose "The Dark Urge" the last free option, which has dark impulses, or simply a custom character.

❖ Historical

It's not very intuitive, but don't take a History giving you the following skills: Deception or/and Performance.

It is better to take Guild Artisan or Street Brat, if you want to avoid unnecessary duplicates at level 3.

❖ Race

Larian Studios has personalized the classes, which has erased the majority of differences since they all have +2 to one characteristic and +1 to another of your choice. You have quite a few viable options for your race, although there are a few that stand out:

- **Half-Elf Wood** : Still a great choice with Darkvision, Stealth proficiency, a movement bonus, and shield handling, among other things.

- **Human** : The handling of polearms, the shield, an additional skill and +25% transported weight.

❖ SKILLS

The Bard literally has access to every skill in the game, you are spoiled for choice. To know what to choose, you need to decide on the composition of your group. If you take a Thief like Astarion, you can do without the associated skills (Deception, Stealth, Sleight of hand). If you invite Gale the Mage, then you can leave him with the skills related to knowledge (Arcana, History, Religion, etc.). Don't forget that you also need skills related to Charisma, both as a Bard and as a group leader.

In the case of the Sword Bard, who normally has a high Dexterity score, it is better to use him to replace the Thief, he is much more competent in this area. Especially since your Intelligence score is pathetic.

So you need:

- Persuasion
- Discretion
- Workaround
- Acrobatics
- Intimidation (optional)

Ironically, you should not take Representation or Deception, since these 2 skills are automatically unlocked at level 3. The interface unfortunately does not warn you.

❖ Features

Considering that you are going to base your character on Dexterity for combat, here is the recommended attribution:

- Force : 10
- Dexterity: 16 to 17
- Constitution : 14
- Intelligence : 8
- Wisdom: 10
- Charisma: 16 to 17

❖ Equipment

There are quite a few viable options depending on your playstyle and the equipment found. Equip light armor that pays off your Dexterity bonus.

As a primary weapon, a longsword or rapier is ideal if you can use them, otherwise a scimitar or shortsword. The important thing is that the weapon used has the "Finesse" attribute. It is advisable to use a shield in your second hand, but playing with two weapons is also an option.

Equip a longbow or crossbow as a ranged weapon, so you always have something to do, even when an enemy is too far away. With your Dexterity score, your DPS will not be too impacted.

❖ Sorts

The Bard remains a spellcaster in his own right, here are the spells recommended for the first levels:

- **Cantrips** : Cruel mockery, Light or Dancing Lights
- **Level 1** : Dissonant Whispers, Thunder Wave, Healing Word, Heroism. Talking with animals can also be useful if you are willing to sacrifice a spell.

❖ Gameplay

As the Bard can do a little bit of everything, his role is also very variable during combat. By default, you will approach enemies, attack them, then use the Sword specialization's mobility ability to move away or move on to the next enemy before attacking again. The Bonus action is used to inspire your teammates, to defend yourself or to cast a quick spell as a healing word.

Other spells are used to kill, weaken or control enemies, or strengthen/heal your allies.

❖ Leveling et dons

It is advisable to choose the Duelist skill at level 3 (it works with a shield) rather than wielding two weapons.

Every 4 levels, you have the choice between a Gift, and +2 Characteristics points to assign (maximum 20 by default). You will have this opportunity 3 times in total, at level 12.

- **Lucky**: Gives the ability to re-roll the dice in many situations, including for your skills. As they are central to the Bard's gameplay, this is a good choice, especially if you don't like reloading the game every time you fail during a dialogue.

- **Robust**: Fighting in melee is dangerous for a Bard, with this Gift you gain more life, which will greatly help you survive.

Paladin Oath of Devotion

For players new to playing Dungeons & Dragons 5th Edition, it can be difficult to guess the best way to optimize each of the 12 classes. It's even more difficult if we take into account the 46 possible subclasses. Here we offer you the main guidelines to follow and the best choices to make in general in Baldur's Gate 3. Nothing prevents you from modifying this build according to the race you wish to play, and the objects available.

It is possible to respecialize your character from a certain point in the adventure, from the camp. Don't be afraid to make mistakes.

❖ *Introducing the Paladin Oath of Devotion*

The Paladin is one of the iconic classes in Dungeons & Dragons, and this is especially true for the Oath of Devotion subclass. He is the archetype of the noble, loyal good holy knight, with his shield and glittering sword. He is a particularly sturdy character, which makes him an excellent tank, but it doesn't stop there, he is more than competent at dealing damage, and he can also heal or assist his teammates. It's also an excellent class choice to act as a group leader, since it has high Charisma and the skills that go with it, such as Persuasion and Intimidation. In summary, he is a very powerful and versatile character. But all this comes at a cost, and if you don't follow the strict rules of the Oath of Devotion (Loyal Good), you'll change your subclass, which will change your powers in ways you don't necessarily want.

+ **Robust and can be used as a tank**

+ **Good single target damage**

+ **Acts as a bonus healer**

+ **High charisma for dialogues and diplomacy**

+ **Suitable for any group**

- **Obligation to make "good" choices under penalty of breaking your oath**

- **Very few area attacks**

- **Few spells**

❖ *Sous-classes*

There are 4 subclasses for the Paladin in Baldur's Gate 3. We'll focus on the Oath of Devotion here, but here's a quick overview of each option:

- **Devotion**: The basic Loyal Holy Paladin.

- Elders: A slightly more natural Paladin, with powers on this theme. Elements, Tanking, and Healing - A little more forgiving of your actions in general.

- **Vengeance** : The vindictive Paladin whose priority is to kill evil beings. It is a much more aggressive subclass, but on the other hand, diplomatic choices are often out of the question.

- **Perjury**: A Paladin who has broken his oath and does not make amends can become a sort of anti-paladin/dark knight, with new evil abilities.

❖ *Origin*

It is strongly recommended that you choose a custom origin, or "Dark Urges" to play this class. This second option is particularly appropriate if you want to play on the duality of the character, with the possibility of resisting his violent impulses, to stay on the path of good.

❖ *Historical*

Choose Soldier (Athletics and Intimidation) or Guild Artisan (Persuasion and Insight). Which are all skills that you will normally choose. Both choices are valid.

❖ *Race*

All races now benefit from the same characteristic bonuses, which greatly reduces disparities. Some remain better than others, depending on masteries and other secondary bonuses.

- **Tieffelin de Zariel** : Darkvision and Fire Resistance. The Thaumaturgy cantrip will also help you intimidate your opponents, if you want to play it. Subsequently, formidable special strikes are unlocked by progressing in level, which is good to take.

- **Drakeids** : An elemental resistance of your choice and a breath of this element. Gold seems most appropriate for a Paladin thematically, as well as for Fire resistance.

- **Nain (Or or Duergar)** : Poison Resistance is good to take, you then have the choice between +1 HP per level, or improved Night Vision and saving throw bonuses. In return, your character is less mobile.

- **Half-orc**: Vision in the dark, a bonus that avoids ending up in agony directly, and more devastating critical hits. This race is more suitable for the Vengeance or Forsworn subclass, but it is still interesting.

❖ SKILLS

With your Origin, you will be able to take everything that is eligible and based on your good Characteristics:

● Athletics

● Intimidation

● Persuasion

● Insight

❖ Features

Your important characteristics are: Charisma, Strength and Constitution. It is recommended to have approximately 16 Charisma, 16 Strength and 14 Constitution (variable depending on the class).

Other characteristics: 8 Intelligence, 10 Charisma. Dexterity is more sensitive, this will influence your armor at the start of the game, as long as you have not found heavy armor, as well as your initiative rolls. We recommend between 8 and 12 Dexterity, depending on the points desired to optimize your main characteristics.

This gives :

● Strength: 16 to 17

● Dexterity: 10

● Constitution: 14 to 16

● Intelligence : 8

● Wisdom: 8 to 10

● Charisma: 16 to 17

❖ Equipment

Equip the heaviest armor you find, a shield, and the best one-handed weapon at your disposal. Equipping a bow or a crossbow is of little use if your Dexterity is low although it will always serve as a backup weapon when no enemy is within range and no spell is worth using. spear.

Subsequently, you can consider using a spear, or even a two-handed polearm, to use the famous Polearm Master (PAM) build often considered meta for the tabletop Paladin.

❖ Sorts

You will only have a few at the start, here is what to select first:

● Forced duel

● Divine Favor

● Divine Punishment

● Wound care

❖ *Gameplay*

The Paladin is a versatile class, the role to play during the battle is quite variable. You will generally position him at the front of the group, and have him rush towards the enemy in order to engage them in melee. This will force them to stay with your tank, otherwise they will suffer attacks of opportunity.

You can configure how your characters' reactions will work in combat. This is particularly important for the Paladin, who has the ability to sacrifice a spell slot to smite the target and inflict holy damage upon hitting them in melee. You can do this on a critical hit, which gives a devastating strike for example.

At the same time, use your spells to heal the group or your Paladin, to strengthen yourself, and to control the enemy.

❖ *Leveling et dons*

Every 4 levels, you will have the choice between +2 characteristic points to freely assign, or to a Gift. This gives a maximum of 3 Gifts or +6 points at level 12 (the maximum). A characteristic cannot exceed 20 by default. Increasing Strength and Charisma is a good idea.

Recommended donations:

● **Polearm Master** : An absolutely monstrous feat that provides an additional attack and the ability to inflict attacks of opportunity in a large area around your character. This will make your Paladin a major obstacle in the path of enemies.

● **Sentinel** : Fantastic with Polearm Master, since it allows you to directly stop enemies, but also prevents them from disengaging.

Forsworn Paladin

With Baldur's Gate 3, Larian Studios made the daring choice to impose real role play limits on the Paladin class, unlike the old titles in the license. It is no longer enough to maintain a high level of reputation, or a certain alignment to maintain this very powerful and prized class. Indeed, depending on your choice of oath, when creating your Paladin, you are forced to follow certain guidelines in your actions. The abuse that the majority of RPG players naturally indulge in, "murder hobo-ism", aka killing and looting everything without getting caught, is not an option, and you really have to be careful about your actions and their consequences repercussions. The good news is that if you want to play a black knight rather than a white knight, you will quickly have the opportunity to change your vocation.

❖ *How to break your oath quickly?*

One of the first opportunities to begin your descent towards the dark side is available at the end of the tutorial, after waking up on the beach. As you advance in the area, you will come across two Tieflings who have imprisoned Lae'Zel. If you don't conduct the negotiations well, or if they fail, you will have to make a painful choice between killing these two innocent idiots, or your potential mate. Killing Tieflings is considered a breach of your oath, which will have repercussions. It must also be said that Lae'Zel is racist, cruel and aggressive at the same time, he is not really a good person to begin with.

Perjured Knight

Then return to the camp, by clicking on the Campfire icon, at the bottom right of the interface. A new NPC will be waiting for you there, the Forsworn Knight, a kind of dark armor. He will explain the situation to you after the betrayal of your oath.

The discussion can be oriented in different ways, and it is even possible to make amends to return to your original oath. But if you persist in this path, you can choose to break your oath completely and become a Forsworn Paladin.

A small cutscene follows, which shows that you are really joining the dark side, like this brave Anakin. The new subclass is then unlocked. Rather than Radiant damage you will inflict Necrotic damage in addition to giving the advantage against the affected target for example. By gaining levels you will unlock new abilities, such as summoning the undead.

Druid Circle of the Moon

For players new to playing Dungeons & Dragons 5th Edition, it can be difficult to guess the best way to optimize each of the 12 classes. It's even more difficult if we take into account the 46 possible subclasses. Here we offer you the main guidelines to follow and the best choices to make in general for the Druid in Baldur's Gate 3. Nothing prevents you from modifying this build according to the race you wish to play, and the objects available.

It is possible to respecialize your character from a certain point in the adventure, from the camp. Don't be afraid to make mistakes.

❖ Presentation of the Druid

Much less popular than other "standard" Dungeons and Dragons classes, the Druid is a spellcaster capable of controlling the surrounding natural energy. He has a strong affinity with nature and its fauna and flora and knows how to benefit from it so much that he can even transform into powerful wild animals to fight, each form offering him serious tactical advantages on the ground.

Since the forms he adopts during combat have hit points distinct from his own, the Druid is inherently a very solid class. However, it remains one of the weakest in the game when it comes to combat. In this guide, we give you the best ideas to make the most of these shapeshifters, thus making your Druid primarily a melee fighter: he therefore takes the place of a "DPS" so to speak, but he is not really designed to "tank" as much damage as a Warrior for example.

+ Lots of different shapeshifters with unique strengths

+ Can heal his allies

+ Extremely versatile, can adapt to almost any situation

+ As a guardian of nature it is good to have him at your side to interact with other Druids, fauna and flora

+ Able to inflict a lot of damage

- Can only wear light or medium armor

- Quite complex to master perfectly, lots of things to do and anticipate

- Quite limited shapeshifting depending on the situation

❖ Druid Subclasses and Combat Style

The Druid has three subclasses unlocked at level 2, which will direct him towards very different roles, depending on your choice. This gives you time to decide which branch to choose:

- **Circle of the Earth** : This is the Druid's "support" specialization. This archetype is generally played at a very distance from enemies and grants many bonuses to your allies in addition to preventing your opponents from moving calmly and inflicting damage using natural energies.

- **Circle of the Moon** : This is the Druid's "melee combat" specialization. Your shapeshifter is greatly enhanced, making the Druid a powerful melee attacker, while allowing him access to a variety of useful spells regardless.

- **Circle of Spores** : This archetype is the equivalent of the Circle of the Earth but "damage" oriented. It focuses on necrotic magic. This is the Druid's "ranged DPS" build with a more necromantic approach than one might imagine!

❖ Origin for the Circle of the Moon Druid

Among the predefined characters available in Baldur's Gate 3, none are Druids. You will later encounter traveling companions who embody this class by default but they cannot be selected as a playable character initially. If you want to embody this class, the option of a personalized character remains accessible!

❖ History for the Circle of the Moon Druid

Without the slightest doubt, the best History for a Druid specialized in Circle of the Moon is Hero of the People. You hone your skills in Survival and Animal Training, two excellent assets that correspond perfectly to the idea we have of a Druid. Your deity will appreciate it! Here are the choices that may be interesting:

- **Acolyte** : You benefit from a bonus to Religion and Insight, two very useful skills during your travels in the mysterious confines full of sometimes divine mysteries that only you can unravel.

- **Stranger**: Very practical especially if you lack knowledge of survival in a hostile environment in your group, the Stranger grants a bonus to Athletics and Survival. A completely viable alternative to the People's Hero.

- **Hero of the people** : Much more focused on what he inspires, the People's Hero benefits from a bonus to Training and Survival. This is the ideal Background for a charismatic leader if that's what you want to embody, in addition to refining your skills in communing with nature.

❖ Race for the Circle of the Moon Druid

Larian Studios has customized the classes, which has erased the majority of differences since they all have +2 to one characteristic and +1 to another of your choice. You have quite a few viable options for your race, although there are a few that stand out:

- **Elf (wood)** : Moves faster, further (Base Racial Speed), can see in the dark (Darkvision), bonus with Short and Long Swords but also Short and Long Bows (Elven Weapon Training) and immunity to magical sleep effects (Fairy Ascendancy)

- **Nain (d'or)** : Moves faster, further (Base Racial Speed), can see in the dark (Darkvision), bonus with Axes, Hatchets, Light Hammers and Warhammers (Dwarven Weapon Training), better resistance with poison effects (Dwarven Resistance) and greater number of hit points (Dwarven Tenacity)

❖ Circle of the Moon Druid Skills

Training, Survival and Perception are excellent statistics since they depend directly on Wisdom, and this is good since it is precisely what we need in Druid. These skills are therefore among the most interesting for this class:

1. **Perception** : A very useful skill that allows you to sometimes detect mysteries in the world even though they are hidden from the eyes of less attentive adventurers. A very good bonus that will serve you permanently.

2. **Dressage** : You are a great friend of animals and can therefore more easily understand what motivates them. Practical to prevent them from taking you for an intruder in their burrow for example!

3. **Survival**: Very effective for following the trail thanks to footprints invisible to the eyes of your allies, or for finding your way in places that ordinary mortals cannot orient themselves effectively. It can also get you out of dangerous interactions that would cost you your life due to lack of knowledge!

❖ Characteristics of the Circle of the Moon Druid

Considering that you will base your character on Wisdom and Constitution for combat, here is the recommended attribution for a Circle of the Moon Druid:

- **Force** : 10
- **Dexterity**: 14
- **Constitution** : 16

- **Intelligence** : 10
- **Wisdom**: 16
- **Charisma**: 8

❖ Equipment for the Circle of the Moon Druid

There are quite a few viable options depending on your playstyle and the equipment found. Equip Heavy Armor to maximize your survival.

As a main weapon, you will mainly seek to obtain a Staff (ideally giving you a new additional spell). A two-handed mace also works well, while a one-handed mace and shield can be useful if you find yourself in trouble in the middle of a melee.

Equip a longbow or crossbow as a ranged weapon so you always have something to do even when an enemy is too far away. Theoretically you shouldn't have any use for it, but you never know, it could save the day!

❖ Circle of the Moon Druid Spells and Abilities

The Druid is largely focused on his ability to cast spells despite the Circle of the Moon specialization. Here are the spells that we recommend you prepare at each level level:

- ◦ *Cantrips*
- Level 1: Magic Club and Thorn Whip
- Level 4: Support
    - ◦ *Prepared Spells*
- Snowat 2: Cure Wounds, Healing Word, Entanglement, Thundering Wave and Ice Knife
- Level 3: Heal Wounds, Healing Word, Entanglement, Thundering Wave, Burning Metal, and Moonbeam
- Level 4: Cure Wounds, Healing Word, Entanglement, Thundering Wave, Burning Metal, Moonbeam, and Flame Blade
- Level 5: Cure Wounds, Healing Word, Entangle, Thunder Wave, Burning Metal, Moonbeam, Fire Blade, Call of Lightning, and Slush Storm
- Level 6: Cure Wounds, Healing Word, Entangle, Thunder Wave, Burning Metal, Moonbeam, Fire Blade, Call of Lightning, Slush Storm, and Partial Restoration
- Level 7: Cure Wounds, Healing Word, Entanglement, Thundering Wave, Burning Metal, Moonbeam, Fireblade, Call of Lightning, Slushstorm, Partial Restoration, and Hailstorm
- Level 8: Cure Wounds, Healing Word, Entangle, Thundering Wave, Burning Metal, Moonbeam, Fireblade, Call Lightning, Slushstorm, Partial Restoration,

Hailstorm, and Summon Sylvan Beings

- Level 9: Cure Wounds, Healing Word, Entangle, Thunder Wave, MBurning Stall, Moonbeam, Fireblade, Call of Lightning, Sleetstorm, Partial Restoration, Hailstorm, Summoning Sylvan Beings, and Contagion

❖ Leveling and Gifts of the Circle of the Moon Druid

It is advisable to choose the Skill Improvement skill at level 4 in order to quickly increase your Wisdom level and gain combat effectiveness.

Every 4 levels, you have the choice between a Gift, and +2 Characteristics points to assign (maximum 20 by default). You will have this opportunity 3 times in total, at level 12.

- Level 4: Skill Improvement (+2 Wisdom)
- Level 8: Resilient (Wisdom)
- Level 12: Tough

Barbarian Berserker

For players new to playing Dungeons & Dragons 5th Edition, it can be difficult to guess the best way to optimize each of the 12 classes. It's even more difficult if we take into account the 46 possible subclasses. Here we offer you the main guidelines to follow and the best choices to make in general for the Barbarian of Baldur's Gate 3. Nothing prevents you from modifying this build according to the race you wish to play, and the objects available.

It is possible to respecialize your character from a certain point in the adventure, from the camp. Don't be afraid to make mistakes.

❖ Presentation of the Barbarian

The Barbarian is the bloodthirsty and reckless version of the Warrior. A formidable melee fighter whose absolute passion is fighting, he clearly takes pleasure in finding himself in the heart of the melee and distributing knuckle salads to his opponents. Expert in melee weapons and close combat, he is mainly considered a "DPS" more than a Tank. His role is to kill, he lives for that.

Very effective and robust even in the heart of the fray, it is often shunned by groups who consider it less versatile than the Warrior... wrongly! Even if he does not have as many assets to help his allies and resistance as the Warrior, the Barbarian is a formidable fighter that we can easily imagine alongside a Druid or a Warrior, precisely. right in the middle of the enemies.

+ Handles almost all weapons perfectly

+ Able to attack multiple times in a single turn

+ Lots of life points

+ Usually wears heavy armor, making him very strong

- **Few spells, if any at all: it focuses on martial skills**

- **Very dependent on the equipment he has at his disposal**

- **Can hardly assist his allies: hitting is his credo**

❖ Origin for the Barbarian Berserker

Among the predefined characters available in Baldur's Gate 3, there is indeed a Warrior. It is Karlach, a Tieffeline with an imposing build and an explosive character who has this honor. Her race and her constitution make Karlach an excellent Barbarian that we highly recommend if you want to opt for a predefined character. Otherwise, the option of a personalized character remains accessible!

❖ History for the Barbarian Berserker

Without a doubt, the best History for a Berserker specialized Barbarian is Soldier. This maximizes his ability to be a frontline fighter while enhancing his proficiency with heavy weapons and armor. Everything we love ! Here are the choices that may be interesting:

● **Soldier:** You benefit from a bonus to Athletics and Intimidation, two skills that are particularly useful in your interactions with the NPCs of the game whether in combat or during your discussions. A real brute!

● **Stranger:** Very practical especially if you lack knowledge of survival in a hostile environment in your group, the Stranger grants a bonus to Athletics and Survival. A completely viable alternative to the Soldier.

❖ Race for the Barbarian Berserker

Larian Studios has customized the classes, which has erased the majority of differences since they all have +2 to one characteristic and +1 to another of your choice. You have quite a few viable options for your race, although there are a few that stand out:

● **Half-Orc** : Moves faster, further (Base racial speed), can see in the dark (Darkvision), has a "cheat death" (Relenty) and increases the damage inflicted by critical hits (Savagery)

● **Dwarf (shield)** : Moves faster, further (Base Racial Speed), can see in the dark (Darkvision), bonus with Axes, Hatchets, Light Hammers and Warhammers (Dwarven Weapon Training), better resistance poison effects (Dwarf Resistance) and mastery of Light and Medium Armor (Dwarf Armor Training)

❖ Barbarian Berserker Skills

The Berserker Barbarian is generally a mobile fighter and quite frightening when seen tumbling onto the field. In any case, this is what he seeks to embody. To achieve this, you must opt   for certain very specific Skills aimed at strengthening these strengths. Here are the three skills that we recommend as a priority:

- **Athletics**: Your Strength allows you to jump higher, further. In addition, this skill allows you to repel your enemies more reliably in addition to preventing them from doing the same to you. A big advantage in a lot of situations.

- **Intimidation** : Obviously this is the perfect skill for the big brute that you play: if nothing helps, you can always threaten your enemies. Threatening them with a good blow of an ax calms most opponents, even the most reckless!

- **Survival**: Very effective for following the trail thanks to footprints invisible to the eyes of your allies, or for finding your way in places that ordinary mortals cannot orient themselves effectively. It can also get you out of dangerous interactions that would cost you your life due to lack of knowledge!

❖ Characteristics of the Barbarian Berserker

Considering that you will base your character on Strength and Constitution for combat, here is the recommended attribution as a Berserker:

- Force : 16
- Dexterity: 14
- Constitution : 16
- Intelligence : 8
- Wisdom: 12
- Charisma: 8

❖ Equipment for the Barbarian Berserker Berserker

There are quite a few viable options depending on your playstyle and the equipment found. Equip Heavy Armor to maximize your survival.

As a primary weapon, any two-handed combat weapon will do, depending on which ones you have access to. Generally, you will mainly want to obtain a Two-Handed Axe, but a Polearm also works very well (it even opens the way to some very powerful possibilities in terms of customization).

Equip a longbow or crossbow as a ranged weapon so you always have something to do even when an enemy is too far away. Despite your low Dexterity score, this remains a significant little extra that can save your entire group in certain situations.

❖ Barbarian Berserker Spells and Abilities

The Barbarian Berserker is not intended to be a great spellcaster, far from it. On the other hand, he has access to many skills that improve his combat effectiveness, and here are the ones you should aim for first:

- **Level 2** : Temerity (obtained by default)
- **Level 3** : Frenzy (obtained by default), Frenzied Attack (obtained by default), Enraged Throw (obtained by default)

- **Level 6** : Additional Rage Charge (obtained by default)

❖ *Leveling a Dons du Barbare Berserker*

It is advisable to choose the Heavy Protection skill at level 4 to allow you to wear heavy armor from the start of your journey.

Every 4 levels, you have the choice between a Gift, and +2 Characteristics points to assign (maximum 20 by default). You will have this opportunity 3 times in total, at level 12.

- **Level 4** : Heavy protection
- **Level 8** : Mastery of heavy armor
- **Level 12** : Polemaster (if you carry a polearm), otherwise Armsmaster

Cleric Domain of Life

For players new to playing Dungeons & Dragons 5th Edition, it can be difficult to guess the best way to optimize each of the 12 classes. It's even more difficult if we take into account the 46 possible subclasses. Here we offer you the main guidelines to follow and the best choices to make in general for the Cleric of Baldur's Gate 3. Nothing prevents you from modifying this build according to the race you wish to play, and the objects available.

It is possible to respecialize your character from a certain point in the adventure, from the camp. Don't be afraid to make mistakes.

❖ *Presentation of the Cleric*

The Cleric is one of the most famous classes in Dungeons and Dragons. A great cleric and true saint among saints, he excels in supporting his allies in multiple ways: from healing to the various reinforcements he provides to weakening enemies, he is a great ally.

Generally made to withstand heavy assaults without really having much offensive capacity, the Cleric adapts perfectly to most group compositions. We sometimes find him replacing a Warrior to "tank", or to act as a healer, or even both at the same time. Clearly not essential, it remains an important element to consider in the composition of your team.

+ **An exceptional support class**

+ **Rather acceptable damage for a "support"**

+ **Can be very durable**

+ **Very versatile: damage, healing, buffs and debuffs at will!**

- **Fairly low mobility**

- **Low or no area damage**

- **Quite complex to master perfectly, lots of things to do and anticipate**

❖ Subclasses and Deities for the Cleric

The Cleric has seven subclasses unlocked at level 3, which will direct him towards slightly different roles, depending on your choice. This gives you time to decide which branch to choose:

● **Domain of Life** : You specialize in treating and supporting your allies as a priority, in addition to mastering Heavy Armor. This is the basic "Healer" archetype that we recommend in this guide.

● **Domain of Light** : Specialization very focused on attacking and applying debuffs to allies mainly. The Cleric's magic damage focused build.

● **Domain of Ruse** : A build focused on deception and illusion, very effective for sneaking up discreetly or changing your appearance to deceive your opponents or interlocutors. Little known but very practical!

● **Domain of Knowledge** : You specialize in knowledge, allowing you to master languages   and have a certain affinity with your interlocutors and control of your allies. This is the build geared towards preventing your opponents from playing while your allies decimate them.

● **Domain of Nature** : It's the druidic version of the Cleric to put it simply. You love and revere animals and Nature, giving you a certain affinity with the brutality of the wild world, giving you a certain natural strength and the ability to communicate with wildlife. You also master heavy armor!

● **Storm Domain** : Oriented towards the magic of Lightning, the Cleric specialized in Domain of the Storm unleashes storms to destroy, push back and throw his adversaries far away, in addition to controlling the mist in order to hide his allies from the eyes of their enemies. You also master War Weapons and Heavy Armor!

● **Domain of War** : A true divine warrior, you embody the divine power in Faerun and fight your enemies with strength and honor under the admiring eyes of your divinity who lavishes you with many blessings to help you. You also master War Weapons and Heavy Armor!

The deity you choose to worship and serve has no direct impact on your character's power. On the other hand, it has certain major implications in the story of the game such as in access to certain elements found here and there during your adventures. Also be aware that NPCs, and even certain deities, do not appreciate the servants of other specific deities. This is particularly the case between the sisters Sélune and Shar who engage in a fierce rivalry: serving one could well inflame the wrath of the other!

❖ Origin for the Cleric Domain of Life

Among the predefined characters available in Baldur's Gate 3, there is indeed a Cleric. It is Shadowheart, a half-elf with a strong character and very specific ambitions who has this honor. Her race and constitution make Shadowheart an excellent Cleric that we highly recommend if you want to opt for a predefined character. Otherwise, the option of a personalized character remains accessible!

- **Note** : By default, Shadowheart is not specialized in Domain of Life. You will need to re-specialize it to gain access to it.

❖ History for the Clerc Domaine de la Vie

Without the slightest doubt, the best History for a Cleric specialized in Domain of Life is Acolyte. You become very perceptive and benefit from serious bonuses to Religion and Insight. Your deity will appreciate it! Here are the choices that may be interesting:

- **Acolyte** : You benefit from a bonus to Religion and Insight, two very useful skills during your travels in the mysterious confines full of sometimes divine mysteries that only you can unravel.

- **Hero of the people** : Much more focused on what he inspires, the People's Hero benefits from a bonus to Training and Survival. This is the ideal Background for a charismatic leader if that's what you want to embody.

❖ Race for the Cleric Domain of Life

Larian Studios has customized the classes, which has erased the majority of differences since they all have +2 to one characteristic and +1 to another of your choice. You have quite a few viable options for your race, although there are a few that stand out:

- **Elf**: Moves faster, further (Base Racial Speed), can see in the dark (Darkvision), bonus with Short and Long Swords but also Short and Long Bows (Elven Weapon Training) and immunity to magical sleep effects (Fairy Ascendancy)

- **Tieffelin** (from Mephistopheles): Moves faster, further (Base Racial Speed), can see in the dark (Darkvision) and halves Fire damage taken (Infernal Resistance) as well as two bonus damage spells at certain levels

❖ Life Domain Cleric Skills

To maximize your effectiveness as a Life Domain Cleric, we recommend that you focus primarily on skills that increase Wisdom. Therefore, we will rather seek to obtain the following skills:

- **Insight**: Very interesting both in your interactions with others and in the wild world, this skill allows you to detect the intentions of your interlocutors and thus avoid being bullied. A must-have in all groups.

- **Perception** : A very useful skill that allows you to sometimes detect mysteries in the world even though they are hidden from the eyes of less attentive adventurers. A very good bonus that will serve you permanently.

- **Medicine**: Sometimes allows you to understand the symptoms from which an unfortunate person you meet is suffering and thus be able to best guide them in their decisions. Not the most useful bonus on a daily basis, but the Wisdom it offers is enough to make it useful!

❖ Characteristics of the Domain of Life Cleric

Considering that you are going to base your character on Wisdom for combat, here is the recommended attribution for the Domain of Life:

- Force : 10
- Dexterity: 14
- Constitution : 16
- Intelligence : 8
- Wisdom: 16
- Charisma: 10

❖ Equipment for the Cleric Domain of Life

There are quite a few viable options depending on your playstyle and the equipment found. Equip Heavy Armor to maximize your survival.

As a main weapon, we mainly recommend a one-handed sword or mace depending on which ones you have access to. Generally, these are the ones that benefit the Cleric the most in terms of statistics granted.

Equip a longbow or crossbow as a ranged weapon so you always have something to do even when an enemy is too far away. Even if you're not there to inflict damage, it's still a nice little extra that can save your entire group in certain situations.

❖ Prepared Spells of the Cleric Domain of Life

The Cleric is largely focused on his ability to cast spells, particularly healing spells with regard to the Domain of Life specialization. Here are the spells that we recommend you prepare at each level level:

  ⋄ *Cantrips*

- Level 1: Holy Flame, Resistance and Assistance
- Level 4: Thaumaturgy

  ⋄ *Prepared Spells*

- Level 2: Tracing Ray, Healing Word, Wound, Injunction and Shield of Faith
- Level 3: Tracing Ray, Healing Word, Wound, Injunction, Shield of Faith and Spiritual Weapon
- Level 4: Tracing Ray, Healing Word, Wound, Injunction, Shield of Faith, Spiritual Weapon, and Bond of Protection
- Level 5: Tracing Ray, Healing Word, Wound, Injunction, Shield of Faith, Spirit Weapon, Bond of Protection, and Group Healing Word

- Level 6: Tracing Ray, Healing Word, Wound, Injunction, Shield of Faith, Spirit Weapon, Bond of Protection, Group Healing Word, and Glyph of Guard

- Level 7: Tracing Ray, Healing Word, Wound, Injunction, Shield of Faith, Spirit Weapon, Bond of Protection, Group Healing Word, Glyph of Guard, and Banishment

- Level 8: Tracing Ray, Healing Word, Wound, Injunction, Shield of Faith, Spirit Weapon, Bond of Protection, Group Healing Word, Glyph of Guard, Banishment and Silence

❖ Leveling and Gifts from the Cleric Domain of Life

It is advisable to choose the Shield Master skill at level 3 to maximize the survival of the Healer in your group (your Cleric therefore).

Every 4 levels, you have the choice between a Gift, and +2 Characteristics points to assign (maximum 20 by default). You will have this opportunity 3 times in total, at level 12.

- Level 4: Shield Master

- Level 8: Tough

- Level 12: Resilient (Wisdom)

Warrior Warmaster

For players new to playing Dungeons & Dragons 5th Edition, it can be difficult to guess the best way to optimize each of the 12 classes. It's even more difficult if we take into account the 46 possible subclasses. Here we offer you the main guidelines to follow and the best choices to make in general for the Warrior of Baldur's Gate 3. Nothing prevents you from modifying this build according to the race you wish to play, and the objects available.

It is possible to respecialize your character from a certain point in the adventure, from the camp. Don't be afraid to make mistakes.

❖ Presentation of the Warrior

A true essential of Dungeons and Dragons, the Warrior is in essence the basic melee fighter of the game. Generally favored by fans of lively decisions and not focused on intense reflections, he is an expert in melee weapons and therefore of close combat. What's more, he can even act as a "Tank" since he perfectly masters wearing heavy armor and has a large number of life points by default.

Very efficient and robust, it is an essential element of many groups. Even if he can be replaced in certain more "exotic" compositions, he is generally said to be a safe bet when it comes to fights and choices focused on violence or robustness. And although he may seem very solitary at first glance, the Warrior brings with him a certain range of skills capable of saving the life of more than one member of the group!

+ Handles almost all weapons perfectly

+ Able to attack multiple times in a single turn

+ Lots of life points

+ Usually wears heavy armor, making him very strong

- Few spells, if any at all: it focuses on martial skills

- Very dependent on the equipment he has at his disposal

❖ Warrior Subclasses and Combat Style

The Warrior has three subclasses unlocked at level 3, which will direct him towards slightly different roles, depending on your choice. This gives you time to decide which branch to choose:

● Champion: Basic archetype of the Warrior, this subclass is based on the Warrior's ability to inflict heavy damage and absorb at least as much. He is a frontline fighter, he leads the assault and usually takes the most damage.

● **Warmaster**: Very versatile, this subclass makes the Warrior capable of having greater control over his opponents using multiple skills, in addition to inflicting significant penalties on them to harm them even more. Less damage, more control!

● **Occult Knight** : Very different from two other subclasses, the Occult Knight combines the martial power of the Warrior with the magical power of an Occultist. This also makes it very versatile, at the expense of a little power in melee, an interesting choice nonetheless.

In terms of combat style, for the Warmaster that we highly recommend in this guide, prefer to move towards two-handed weapons for a more offensive style of play, or towards Protection for gameplay more focused on the assistance of your allies and the possibility of taking a lot of damage.

❖ Origin for the Warmaster Warrior

Among the predefined characters available in Baldur's Gate 3, there is indeed a Warrior. It's Lae'zel, a Githyanki with a strong character and very specific ambitions who has this honor. Her race and her constitution make Lae'zel an excellent Warrior who we highly recommend if you want to opt for a predefined character. Otherwise, the option of a personalized character remains accessible!

❖ History for the Warmaster Warrior

Without the slightest doubt, the best History for a Warmaster specialized Warrior is Soldier. This maximizes his ability to be a frontline fighter while enhancing his proficiency with heavy weapons and armor. Everything we love ! Here are the choices that may be interesting:

● **Soldier**: You benefit from a bonus to Athletics and Intimidation, two skills that are particularly useful in your interactions with the NPCs of the game whether in combat or during your discussions. A real brute!

- **Stranger**: Very practical especially if you lack knowledge of survival in a hostile environment in your group, the Stranger grants a bonus to Athletics and Survival. A completely viable alternative to the Soldier.

- **Hero of the people**: Much more focused on what he inspires, the People's Hero benefits from a bonus to Training and Survival. This is the ideal Background for a charismatic leader if that's what you want to embody.

❖ Race for the Warmaster Warrior

Larian Studios has customized the classes, which has erased the majority of differences since they all have +2 to one characteristic and +1 to another of your choice. You have quite a few viable options for your race, although there are a few that stand out:

- **Half-Orc** : Moves faster, further (Base racial speed), can see in the dark (Darkvision), has a "cheat death" (Relenty) and increases the damage inflicted by critical hits (Savagery)

- **Tieffelin** : Moves faster, further (Base Racial Speed), can see in the dark (Darkvision) and halves Fire damage taken (Infernal Resistance)

❖ Warrior Skills Warmaster

The Warmaster Warrior is generally a mobile fighter and quite frightening when seen tumbling onto the field. In any case, this is what he seeks to embody. To achieve this, you must opt   for certain very specific Skills aimed at strengthening these strengths. Here are the three skills that we recommend as a priority:

- **Athletics**: Your Strength allows you to jump higher, further. In addition, this skill allows you to repel your enemies more reliably in addition to preventing them from doing the same to you. A big advantage in a lot of situations.

- **Persuasion** : Charisma skill, this allows you to persuade your interlocutors to act in a way different from the behavior they would normally have.

- **Intimidation** : Charisma Skill, obviously this is the perfect skill for the big brute that you play: if nothing helps, you can always threaten your enemies. Threatening them with a good blow of an ax calms most opponents, even the most reckless!

❖ Characteristics of the Warmaster Warrior

Considering that you will base your character on Strength and Constitution for combat, here is the recommended attribution as a War Master:

- Force : 16
- Dexterity: 14
- Constitution : 16

- Intelligence : 8
- Wisdom: 10
- Charisma: 10

❖ Equipment for the Warmaster Warrior

There are quite a few viable options depending on your playstyle and the equipment found. Equip Heavy Armor to maximize your survival.

As a primary weapon, any two-handed combat weapon will do, depending on which ones you have access to. Generally, you will mainly be looking to obtain a Greatsword or a Two-handed Axe, but a Polearm also works very well (it even opens the way to some very powerful possibilities in terms of customization).

Equip a longbow or crossbow as a ranged weapon so you always have something to do even when an enemy is too far away. Despite your low Dexterity score, this remains a significant little extra that can save your entire group in certain situations.

❖ Warrior Warmaster Spells and Skills

The Warmaster Warrior is not intended to be a great spellcaster, far from it. On the other hand, he has access to many skills that improve his combat effectiveness, and here are the ones you should aim for first:

- Level 1: Second Wind (obtained by default)
- Level 2: Passion (obtained by default)
- Level 3: Defensive Footwork, Maneuvering Attack, Riposte
- Level 5: Additional attack (obtained by default)
- Level 7: Trip, Sweep

❖ Leveling and Gifts of the Warrior Warmaster

It is advisable to choose the Weapon Master skill at level 4 in order to maximize the bonuses you get from your weapons, thus greatly improving your combat effectiveness (the basis for a Warrior).

Every 4 levels, you have the choice between a Gift, and +2 Characteristics points to assign (maximum 20 by default). You will have this opportunity 3 times in total, at level 12.

- Level 4: Weapons Master
- Level 8: Heavy Armor Proficiency
- Level 12: Polemaster (if you carry a polearm), otherwise Gaillard

Monk Path of Shadow

For players new to playing Dungeons & Dragons 5th Edition, it can be difficult to guess the best way to optimize each of the 12 classes. It's even more difficult if we take into account the 46 possible subclasses. Here we offer you the main guidelines to follow and the best choices to make in general for the Monk of Baldur's Gate 3. Nothing prevents you from modifying this build according to the race you wish to play, and the objects available.

It is possible to respecialize your character from a certain point in the adventure, from the camp. Don't be afraid to make mistakes.

❖ Presentation of the Monk

If there is a class that is not very popular in Dungeons and Dragons, and even more so in Baldur's Gates 3, it is the Monk. The reason for this is simple and twofold: on the one hand, it is not a class commonly judged as suited to the "Heroic Fantasy" genre (wrongly), and on the other hand its very particular way of tackling the fights makes it unique in its kind, and therefore off-putting at first glance.

The Monk does not in fact need to equip armor to be viable, on the contrary he seeks to emancipate himself from it as much as possible. Likewise, the weapons he can equip are very limited since it is his unarmed combat or through weapons considered "Monk" that he favors. And if at first glance this may seem surprising or even weak, he positions himself quite quickly (around level 5) as a real brute in melee combat. He is often given the same role as a Ranger or Rogue.

+ **Can turn invisible in the middle of combat**

+ **Large bonuses granted by invisibility and camouflage**

+ **Unrivaled mobility, can even teleport and jump very far**

+ **Able to deflect projectiles (including magic)**

+ **Has the same role as a Rogue or Ranger in terms of disarming traps and picking**

+ **Difficult to knock down (Athletics)**

+ **The armor he does not equip can be used by another member of the group!**

- **Quite weak until level 5**

- **Should not equip armor**

- **Fairly limited exploitable weapons**

- **Can seem off-putting at first if you don't give it a chance**

❖ Subclasses for the Monk

The Monk has three subclasses unlocked at level 3, which will direct him towards very different roles, depending on your choice. This gives you time to decide which branch to choose:

● Path of the Four Elements: Much more dependent on spells than his Palm and Shadow counterparts, this Monk unleashes the elements through powerful techniques combining martial and elemental power for a barrage of explosive blows.

- Way of the Palm: Exclusively adept at unarmed combat, this Monk unleashes the martial skills at his disposal by delivering well-placed blows to his opponents in order to knock them out without resorting to force of arms.

- Shadow Path: Mastering the art of disappearance and concealment, he emerges from the shadows to deal deadly blows to his opponents before disappearing in a screen of smoke until his next attack. Plays more or less like a Rogue Assassin.

❖ Origin for the Shadowpath Monk

Among the predefined characters available in Baldur's Gate 3, none are Monks. You will later encounter traveling companions who embody this class by default but they cannot be selected as playable characters initially. If you want to embody this class, the option of a personalized character remains accessible!

❖ History for the Shadowpath Monk

Without a doubt, the best History for a Monk specialized in Shadow Path is Street Kid. This maximizes his ability to unlock chests and disarm traps primarily while greatly improving his Dexterity. It's extremely powerful. Here are the choices that may be interesting:

- **Street kid** : Greatly improves your Sleight of Hand and Stealth skills while strengthening your Dexterity. This is one of the best choices in Ranger since it is everything we expect from this class.

- **Criminal**: Strengthens your skills in Deception and Stealth, two particularly important elements as a Monk. Even if Dexterity increases less with this history, it is still an excellent thing to gain all these characteristics which will be useful to you very often with this class.

- **Charlatan** : Quite similar to Criminal in terms of stat gains, this background improves your ability to fool people with Deception before robbing them with Sleight of hand. About as good as the Criminal, but the playstyle supposedly expected of a Charlatan is much more vicious, so to speak.

❖ Race for the Shadowpath Monk

Larian Studios has customized the classes, which has erased the majority of differences since they all have +2 to one characteristic and +1 to another of your choice. You have quite a few viable options for your race, although there are a few that stand out:

- **Elf**(Woods): Moves faster, further (Base racial speed), can see in the dark (Darkvision), bonus with Short and Long Swords but also Short and Long Bows (Elven Weapon Training) and immunity to magical sleep effects (Fairy Ascendancy)

- **Tieffelin** (from Asmodeus): Moves faster, further (Base Racial Speed), can see in the dark (Darkvision), halves Fire damage taken (Infernal Resistance), and gains the Production cantrip of flame

❖ Shadowpath Monk Skills

For a Shadowpath Monk, you will generally be looking to increase your Dexterity-enhancing skills. So aim for those useful in terms of discretion, discovery of secret elements and survival in a hostile environment. We therefore recommend the following skills:

● **Discretion**: You move more discreetly and avoid being spotted by surrounding NPCs. This gives you a serious tactical advantage during combat and also allows you to try to pickpocket items from the pockets of both your enemies and your allies... At your own risk that being said.

● **Acrobatics**: You are more resistant to effects that push you back and knock you down, an important element when you find yourself in melee and which can have serious repercussions if the opponent gets his way!

● **Workaround**: Greatly strengthens your skills in lockpicking of all kinds, disarming traps and pickpocketing.

❖ Shadowpath Monk Traits

Considering that you will base your character on Strength and Constitution for combat, here is the recommended attribution as a follower of the Path of Shadow:

● Force : 10

● Dexterity: 17

● Constitution : 14

● Intelligence : 8

● Wisdom: 16

● Charisma: 8

❖ Shadowpath Monk Equipment

Unlike all the other classes in the game, the Monk's main characteristic is that it does not need armor to defend itself. On the contrary in fact, thanks to the Defense without armor passive your Wisdom modifier is added to your armor class provided that you are not wearing armor. A very powerful and interesting bonus that you must exploit as much as possible at all costs.

● **Important** : All pieces indicated with a type of armor at the bottom of their description are considered "Armor". If nothing is indicated on a given piece you can equip it as a Monk, if the indication "Light armor", "Intermediate armor" or "Heavy armor" is present then it should be banned.

All other equipment slots can be exploited as a Monk: Cloak, Amulet, Rings and obviously underwear are tolerated. Take advantage of it by equipping pieces that improve your melee effectiveness as much as possible.

In terms of weapons, the Monk is again very particular since he only benefits from certain very specific weapons considered as "Monk weapons". In this category, we in fact include all so-called "martial" and "simple" weapons:

- Longsword
- Rapier
- Cimeterre
- Battle Ax
- Morgenstern
- Short sword
- Trident
- war hammer
- War Spear
- Club
- Dagger
- Javelin
- Light hammer
- One-handed mace
- Stick
- Faux
- Lance

To benefit from the bonuses that your class gives you as soon as you fight with your bare hands or with a "Monk weapon", you should therefore generally avoid two-handed weapons (with the exception of sticks). Likewise, any weapons you're not proficient with should be banned, and you'll generally be looking for powerful bonuses through your combat implements rather than actual damage (you deal a lot of it anyway). by default with your little hands).

❖ Shadowpath Monk Spells and Abilities

   ❖ Cantrips

- Level 3: Minor Illusion (obtained by default)

   ❖ Skills and Actions

- Level 1: Flurry of Blows (obtained by default)
- Level 3: Shadow Artist: Hide, Shadow Artist; Traceless Passage, Shadow Artist Darkness, Shadow Artist: Darkvision, Shadow Artist: Silence (obtained by default)

198

- Level 5: Stunning Strike (melee), Stunning Strike (unarmed), Shadow Shroud (obtained by default)
- Level 6: Shadow Stride (obtained by default)
- Level 11: Shadow Strike, Shadow Strike: Unarmed Attack (obtained by default)

❖ Leveling and Gifts of the Shadow Path Monk

It is advisable to choose the Athlete skill at level 3 in order to increase your combat skills using Dexterity in addition to allowing you more fluidity in your movements.

Every 4 levels, you have the choice between a Gift, and +2 Characteristics points to assign (maximum 20 by default). You will have this opportunity 3 times in total, at level 12.

- Level 4: Athlete (+1 Dexterity)
- Level 8: Improved characteristics (+2 Constitution)
- Level 12: Improved characteristics (+2 Constitution)

Hunter Hunter

For players new to playing Dungeons & Dragons 5th Edition, it can be difficult to guess the best way to optimize each of the 12 classes. It's even more difficult if we take into account the 46 possible subclasses. Here we offer you the main guidelines to follow and the best choices to make in general for the Baldur's Gate 3 Ranger. Nothing prevents you from modifying this build according to the race you wish to play, and the items available.

It is possible to respecialize your character from a certain point in the adventure, from the camp. Don't be afraid to make mistakes.

❖ Presentation of the Ranger

More discreet in its approach to combat and its popularity among Dungeons and Dragons players, the Ranger is a very particular class. Indeed, he is an excellent tracker gifted with a certain mastery of survival in hostile environments and difficult terrain, he therefore benefits from positioning himself meticulously before unleashing all his power on his poor prey. He is generally known for his great mastery of ranged weapons, the Longbow in particular, but he is relatively brilliant in melee as well.

The Ranger has the role of what one would easily call a thief in a group. He is not there to take a lot of damage but to inflict as much as possible: he is the one who takes the opponent by surprise, never the other way around. Exclusively "DPS" oriented, he is equipped with an arsenal capable of putting frightening pressure on a group of creatures in addition to being able to help his allies using multiple skills giving them advantage on the field.

+ **Usually played at long range**

+ **Lots of skills to give the group an advantage**

+ Able to inflict a lot of damage

+ Mobility that is difficult to match

- Almost no spells to speak of

- Requires careful placement to play correctly

- Fewer life points than common classes

❖ Subclasses, Combat Style, Defensive Tactics, Prey and Nemesis for the Ranger Hunter

The Ranger has three subclasses unlocked at level 3, which will direct him towards very different roles, depending on your choice. This gives you time to decide which branch to choose:

● **Beast Master** : This is the archetype that uses a familiar that it can summon once per day to assist it in combat. Quite weak, it nevertheless allows you to have an allied creature to provide a diversion while you sneak up behind your enemies.

● **Dark Tracker** : As powerful as it is difficult to play, this is the build that focuses on surprise attacks to exterminate your enemies in one or two hits at most. Positioning is key and is by far the biggest liability of this build.

● **Hunter** : Grand master of ranged weapons, the Hunter benefits greatly from elevated positions in order to bombard his enemies from strategic points. He is vulnerable alone but formidable in groups.

In terms of combat style, for the Hunter who we highly recommend in this guide, prefer to move towards Archery for a style of play exclusively dedicated to long-range attacks (regardless of the weapon used).

As for sworn enemies, you can select a maximum of two, so we recommend the following:

● **Level 1**: Bounty Hunter + Urban Tracker

● **Level 6**: Mage Breaker + Wild Beast Tamer

You can also choose up to two types of preferred prey, and here are the ones we recommend:

● **Level 3** : Colossus Slayer

Finally, you have the choice between several Defensive Tactics, and we recommend the following:

● **Level 7** : Morale of steel

❖ Origin for the Ranger Hunter

Among the predefined characters available in Baldur's Gate 3, none are Rangers. You will later encounter traveling companions who embody this class by default but they cannot be selected as a playable character initially. If you want to embody this class, the option of a personalized character remains accessible!

❖ History for the Ranger Hunter

Without a doubt, the best History for a specialized Hunter Hunter is Street Kid. This maximizes his ability to unlock chests and disarm traps primarily while greatly improving his Dexterity. It's extremely powerful. Here are the choices that may be interesting:

- **Street kid**: Greatly improves your Sleight of Hand and Stealth skills while strengthening your Dexterity. This is one of the best choices in Ranger since it is everything we expect from this class.

- **Criminal** : Strengthens your skills in Deception and Stealth, two particularly important elements as a Ranger. Even if Dexterity increases less with this history, it is still an excellent thing to gain all these characteristics which will be useful to you very often (depending on your playing style).

- **Charlatan** : Quite similar to Criminal in terms of stat gains, this background improves your ability to fool people with Deception before robbing them with Sleight of hand. About as good as the Criminal, but the playstyle supposedly expected of a Charlatan is much more vicious, so to speak.

❖ Breed for the Ranger Hunter

Larian Studios has customized the classes, which has erased the majority of differences since they all have +2 to one characteristic and +1 to another of your choice. You have quite a few viable options for your race, although there are a few that stand out:

- **Elf**(Woods): Moves faster, further (Base racial speed), can see in the dark (Darkvision), bonus with Short and Long Swords but also Short and Long Bows (Elven Weapon Training) and immunity to magical sleep effects (Fairy Ascendancy)

- **Drow** : Can see in the dark (Darkvision), bonuses with Short Swords, Hand Crossbows and Rapiers (Drow Weapons Training) and immunity to magical sleep effects (Fairy Ascendancy)

❖ Hunter Ranger Skills

For a Ranger Hunter, you will generally seek to increase your skills in stealth, discovery of secret elements and survival in hostile environments. We therefore recommend the following skills:

1. **Discretion**: You move more discreetly and avoid being spotted by surrounding NPCs. This gives you a serious tactical advantage during combat and also allows you to try to pickpocket items from the pockets of both your enemies and your allies... At your own risk that being said.

2. **Insight**: Very interesting both in your interactions with others and in the wild world, this skill allows you to detect the intentions of your interlocutors and

thus avoid being bullied. A must-have in all groups.

3. **Survival**: Very effective for following the trail thanks to footprints invisible to the eyes of your allies, or for finding your way in places that ordinary mortals cannot orient themselves effectively. It can also get you out of dangerous interactions that would cost you your life due to lack of knowledge!

❖ Characteristics of the Ranger Hunter

Considering that you will base your character on Dexterity and Constitution for combat, here is the recommended attribution for a Hunter:

- Force : 10
- Dexterity: 16
- Constitution : 16
- Intelligence : 8
- Wisdom: 14
- Charisma: 10

❖ Equipment for the Ranger Hunter

There are quite a few viable options depending on your playstyle and the equipment found. Equip Medium Armor to maximize your survival despite the low potential of the armor you can wear.

As a primary weapon, any two-handed combat weapon will do, depending on those you have access to and the advantages your race gives you. Generally, you'll mostly be looking to get a Two-Handed Sword or two Hatchets, but exceptions can always happen.

Equip a Longbow or a Heavy Crossbow as a ranged weapon since it is this weapon that will inflict the overwhelming majority of your damage. The only exception concerns the Drow who will benefit from carrying two light crossbows (one in each hand) since their racial gives them mastery of this type of weapon.

❖ Spells and skills of the Ranger Hunter

- The Ranger Hunter is not intended to be a great spellcaster, far from it. On the other hand, he has access to many skills that improve his combat effectiveness, and here are the ones you should aim for first:
- Level 2: Hunter's Mark and Trick Strike
- Level 3: Fog patch
- Level 5: Thorn Growth
- Level 6: Silence
- Level 9: Lightning Arrow

❖ Leveling and Gifts of the Ranger Hunter

It is advisable to choose the Marksman skill at level 4 to maximize your ranged damage potential.

Every 4 levels, you have the choice between a Gift, and +2 Characteristics points to assign (maximum 20 by default). You will have this opportunity 3 times in total, at level 12.

- **Level 4** : Sniper
- **Level 8**: Master of arms
- **Level 12** : Master Crossbowman (if you carry a crossbow), otherwise Martial Savagery

Rogue Assassin

❖ Presentation of the Dodger

The Rogue is par excellence our idea of   a thief: sneaky, mischievous and unpredictable. Very popular with some Dungeons and Dragons players, this is a class that shines for the cunning it uses to get around sometimes colossal problems He is an unrivaled master of infiltration and assassination without leaving the slightest trace.

Unfortunately, in Baldur's Gate 3 the Dodger suffers greatly. The game is in fact not really suited to a class whose basic principle is to act in the shadows, out of sight, and even if it can offer you certain advantages it remains a serious handicap in the most group compositions... Unless you opt for a narrative more oriented towards pickpocketing, murder and infiltration, in which case it will undoubtedly shine.

Despite its obvious and glaring flaws in the game, the Dodger remains a very fun and interesting class to play, offering a unique perspective in how players approach combat as well as encounters. And if normally, at least at the beginning, you could play without thinking too much, with a Dodger in the group you will have to demonstrate tactics, patience and ingenuity to get the most out of his talents .

+ Very discreet, almost invisible as soon as it is hidden

+ Grand master of Sleight of hand, Discretion and even Deception

+ Can deal a lot of damage if given the chance

+ Designed to infiltrate prohibited places

+ Can take the appearance of other individuals

+ Very useful outside of combat mainly (dialogues, exploration...)

- One and only active skill, very restrictive in addition

- By far the most difficult class to play in Baldur's Gate 3, requires increased mastery of positioning and the game as a whole to be played correctly

- Very few hit points, even at high levels

- Overall quite weak, and almost unplayable if he does not have the advantage in combat

❖ Subclasses for the Dodger

The Rogue has three subclasses unlocked at level 3, which will direct him towards very different roles, depending on your choice. This gives you time to decide which branch to choose:

- **Thief**: Very largely focused on his abilities to blend into the background and plunder resources to which he should not have access, the Thief is very good at infiltrating and pickpocketing, or even outright stealing entire treasures. We also know he has a certain talent for lockpicking, obviously, and his movements are particularly fast in addition to having access to several bonus actions each turn.

- **Arcane Swindler** : Very particular, the Arcane Swindler combines the strengths of the Dodger and the Magician. Able to camouflage himself like his counterparts, he also knows how to cast spells and does just about everything in reality. His great versatility combined with the obvious advantages of the two classes in one that he embodies make him an asset of choice in combat as well as in your interactions.

- **Assassin** : The Assassin is a silent murderer who does not hesitate to use the worst deviousness to defeat his adversaries. Very good at duels and assassinating single targets in general, he shines when he has an unfair advantage over his victim. Chivalry, we forget!

❖ Origin for the Rogue Assassin

Among the predefined characters available in Baldur's Gate 3, there is indeed a Rogue. It is Astarion, a high elf, a smooth talker and adept at all pleasures, who has this honor. His race and constitution make Astarion an excellent Rogue that we highly recommend if you want to opt for a predefined character. Otherwise, the option of a personalized character remains accessible!

❖ History for the Rogue Assassin

Without the slightest doubt, the best History for a Rogue specialized in Assassin is Criminal. This maximizes his ability to manipulate his interlocutors by mainly lying to them while greatly improving his Dexterity and his ability to move in the shadows. It's extremely powerful. Here are the choices that may be interesting:

- **Criminal** : Strengthens your skills in Deception and Stealth, two particularly important elements as a Ranger. Even if Dexterity increases less with this history, it is still an excellent thing to gain all these characteristics which will be useful to you very often (depending on your playing style).

- **Street kid** : Greatly improves your Sleight of Hand and Stealth skills while strengthening your Dexterity. This is one of the best choices in Ranger since it is everything we expect from this class.

- **Charlatan** : Quite similar to Criminal in terms of stat gains, this background improves your ability to fool people with Deception before robbing them with Sleight of hand. About as good as the Criminal, but the playstyle supposedly expected of a Charlatan is much more vicious, so to speak.

❖ *Race for the Rogue Assassin*

Larian Studios has customized the classes, which has erased the majority of differences since they all have +2 to one characteristic and +1 to another of your choice. You have quite a few viable options for your race, although there are a few that stand out:

- **Elf**(Woods): Moves faster, further (Base racial speed), can see in the dark (Darkvision), bonus with Short and Long Swords but also Short and Long Bows (Elven Weapon Training) and immunity to magical sleep effects (Fairy Ascendancy)

- **Drow** : Can see in the dark (Darkvision), bonuses with Short Swords, Hand Crossbows and Rapiers (Drow Weapons Training) and immunity to magical sleep effects (Fairy Ascendancy)

❖ *Rogue Assassin Skills*

For a Rogue Assassin, you will generally seek to increase your skills in stealth, discovery of secret elements and survival in hostile environments. We therefore recommend the following skills:

1. **Discretion**: You move more discreetly and avoid being spotted by surrounding NPCs. This gives you a serious tactical advantage during combat and also allows you to try to pickpocket items from the pockets of both your enemies and your allies... At your own risk that being said.

2. **Insight**: Very interesting both in your interactions with others and in the wild world, this skill allows you to detect the intentions of your interlocutors and thus avoid being bullied. A must-have in all groups.

3. **Survival**: Very effective for following the trail thanks to footprints invisible to the eyes of your allies, or for finding your way in places that ordinary mortals cannot orient themselves effectively. It can also get you out of dangerous interactions that would cost you your life due to lack of knowledge!

❖ *Characteristics of the Rogue Assassin*

Considering that you will base your character on Dexterity for combat, and roughly balance the other statistics with the exception of Strength and Intelligence, here is the recommended attribution for an Assassin:

- Force : 9
- Dexterity: 16

- Constitution : 16
- Intelligence : 8
- Wisdom: 14
- Charisma: 14

❖ Equipment for the Rogue Assassin

There are quite a few viable options depending on your playstyle and the equipment found. Equip Light Armor to maximize your survival despite the low potential of the armor you can wear.

As for weapons, since you are supposed to play with the Gift: Two-Weapon Fighter then we strongly recommend that you opt for weapons adapted to your race (those which inflict the most damage too). Generally, for a Drow we will seek to obtain Rapiers where a Wood Elf will mainly favor short and long Swords for example.

Equip a Longbow or Heavy Crossbow as a ranged weapon. Even if you're not supposed to be a master at dealing damage from a distance, it can save you from a very dangerous situation of shooting an arrow at a target in the distance in certain situations!

❖ Rogue Assassin Spells and Skills

The Rogue Assassin is very particular in his genre since he only has one and only one skill by default: Sneak attack. His entire style of play is based on this, but also and above all on your ability to have the advantage over your opponents by remaining camouflaged as much as possible and taking them by surprise.

The Dodger Assassin's play style is unfortunately not optimal in Baldur's Gate 3, and even if it benefits from a certain interest in certain specific situations, it remains a fairly weak class compared to the others available to players. This does not mean that it is not playable or that you should not play it: with a little bit of talent and ingenuity you will definitely be able to make your Dodger Assassin shine that's a certainty !

❖ Leveling and Gifts of the Rogue Assassin

It is recommended to choose the Two-Weapon Fighter skill at level 4 to improve your melee survival while allowing you to wield two non-light weapons at once.

Every few levels, you have the choice between a Gift, and +2 Characteristics points to assign (maximum 20 by default). You will have this opportunity 3 times in total, at level 12.

- Level 4: Two-Weapon Fighter
- Level 8: Vigilant
- Level 10: Improved characteristics (+2 Dexterity)
- Level 12: Tough

ROMANCES AND COMPANIONS

ORIGINAL COMPANIONS & CHARACTERS

As the introductory cinematic and gameplay video of Baldur's Gate 3 revealed to us various very unlucky prisoners are in the hands (or tentacles) of the Illithids. By launching a new game, as in the old Baldur's Gate, you are offered the choice of fully configuring a character, with their gender, race, class, history, etc. This is not what we are going to focus on here, but on the second option, that of choosing one of the 7 so-called original people, as well as the other companions who can join the adventure later.

❖ What is an original character?

With a fixed appearance and race, these colorful characters have a very particular, if not complex, history and circumstances. These characters have special lines and events, as well as a special questline linked to their personal story. After creating or choosing your character (original or not), the remaining original characters can be encountered in-game and recruited into your group if you do it well. They then act as NPCs and companions for your group. When recruiting them, you will not be able to choose their class or their characteristics, but it is possible to respecialize them at the camp by speaking to Blight.

To give you a comparison, it's as if you could choose to play Jaheira, Imoen, Edwin or Minsc in the first Baldur's Gate for example.

On the one hand, the original characters severely restrict your creative freedom, but on the other, they allow you to launch the game directly without any hassle. In Divinity 2, the developers were criticized for seriously penalizing non-original characters, since they were deprived of a quest series linked to their history. According to the Larian studio, this should no longer be a problem, although this remains to be discovered.

❖ Original characters

Here is the known information on the 7 original characters / companions present in the game. Let us point out in passing that they can absolutely leave your group or even attack you if they do not appreciate your actions. Their personal story can take different turns depending on your actions and circumstances. All these characters have in common that they have been kidnapped by the Illithid and have been parasitized by a special larva which risks devouring their brains if they do not act quickly.

❖ Astarion

Astarion is a rogue high elf transformed into a vampire, condemned to serve a cruel master for centuries. Strangely, the insertion of the illithid tadpole gave it the ability to be able to live in broad daylight and expose itself to sunlight without undergoing spontaneous combustion. His story seems to revolve around the hiding of his true identity, his thirst for blood and the turn his relationship will take with his vampire master.

- ### Ombrecoeur (Shadowheart)

A half-high elf cleric of the domain of deception, and priestess of Shar, the evil goddess of darkness. Shadowheart was sent on a suicide mission to steal an item of great power. On the one hand, she questions her own faith, and on the other, she seems to have strange and uncontrolled magic. In any case, she has enemies on all sides, as well as an ancient secret to discover.

- ### Gayle (Gale)

This human magician has only one ambition: to become the greatest mage Faerun has ever known in order to charm the goddess of magic. But his thirst for magic has led to disaster, and a Netheril Orb of Destruction now pulses within his chest, with a countdown that will unleash an explosion capable of wiping out an entire city. Gayle is confident in his ability to overcome this ordeal, but time is not on his side. You will need to regularly destroy powerful magic items to power the orb, in order to save time.

- ### Lae'zel

This experienced Githyanki warrior is fearsome and fierce, even by the standards of her race. As she runs the risk of transforming into one of the monsters she swore to destroy, Lae'zel must prove that she is worthy of returning to her people, if they do not execute her first.

- ### Wild

Wyll is a human occultist who was born into a noble family and later earned a heroic name as the "Blade of the Borders". But he keeps the fact of having made a pact with the devil well hidden, and he is desperate to escape this hellish contract, even if it means rescuing the seductive creature with whom he was signed.

- ### Karlach

Karlach is a tiefling barbarian who has just escaped the charred battlefields of the Blood War. Her violent past is visible on her body, in the form of numerous scars, burns and other tattoos, the marks of a legendary warrior. After managing to escape the eternal war between devils and demons, she is determined to enjoy every moment of her new existence. At least, as much as she can, given the infernal machine that serves as her heart. This gift from the archdevil in whose service she was certainly makes her formidable, but when her emotions take over, she destroys everything she touches, including you.

- ### The Dark Urge

While the first 6 original characters are fixed, the last one is fully customizable. He can be of any race or class. However, it has a story linked to its origin, from Dark

Impulsions. His default race is a pale-skinned, red-eyed Drakeid, and he is meant to represent the darkest possible side of the game in terms of morality and actions. He is encouraged to behave in the most cruel and violent manner possible under all circumstances. You will be free to try to resist, or on the contrary, to give free rein to his bloodlust. The insertion of the Illithid tadpole had the side effect of erasing his memory, and now all he has left are his dark urges. This character also has a unique servant, Sceleritas Fel, who turns out to be as vile as he is loyal, and who wants his master to find his place. You will have to discover the mysteries of his past, and choose which links to establish with your other companions. Becoming friends and developing romance is still possible. Unlike other original characters, "The Dark Urge" does not exist in the form of a companion to recruit, it is reserved for the protagonist.

❖ Normal companions, to be recruited during the adventure (non-origin)

A few conventional companions can be recruited during the adventure if you meet the requirements for this. You will usually need to complete a special quest. Some are mutually exclusive.

⬦ Halsin

This particularly muscular Wood Elf Druid is the leader of the Druid Sanctuary in the first act of the game. Once freed from his cage in the goblin camp, he can join your group temporarily, but you will have to lift the curse of the act 2 in order to make it a permanent member. Halsin's specialty is the transformation into a bear, formidable in melee.

⬦ Minthara

This Drow from Lloth Paladine is one of those responsible for the goblin attack on the druid sanctuary. Like the protagonist and the original characters, she also has an Illithid larva in her head, which gives her special powers. Minthara asks the player and their party to find the shrine, then help attack it, as well as destroy it. Collaborating with her misdeeds makes it possible to develop a relationship with her, and to make her join the team, but this is apparently not compatible with the recruitment of Halsin.

⬦ Jaheira

Jaheira is one of the hero's companions in Baldur's Gate 1, 2, and Throne of Bhaal. She is a half-wood elf druidess. Her race explains the fact that she is still alive more than a century later, but the years have left their mark. She is encountered during chapter 2, and can be recruited at the end of it, at least, if you make the right choices regarding Chantenuit.

⬦ Minsk

Minsc is a kind-hearted, but rather mentally limited, human ranger who has had many adventures on the Sword Coast and beyond. He joined many adventuring parties throughout the 14th and 15th centuries, including the Heroes of Baldur's Gate. Minsc was famous for his habit of talking to a hamster called Boo, which he believes to be a miniature giant space hamster, although no one else has been able to confirm this. He accompanied the protagonist in Baldur's Gate 1, 2 and Throne of Bhaal, but he was petrified shortly after with Boo. His petrification was lifted recently, which explains why he is still alive, despite the time that has passed between games. His recruitment in Baldur's Gate 3 proves difficult, it is only possible during chapter 3, and only by having also recruited Jaheira.

- *Mercenaries*

After meeting Withered in the damp crypt, he will join your camp and offer you his services. One of them involves recruiting mercenaries in exchange for 200 gold coins. To put it simply, this allows you to recruit generic companions, devoid of personality, but which you can configure freely: race, class, characteristic. This is a particularly useful option if you have decided to kill off some potential mates, or your relationship has deteriorated too much.

ROMANCES BETWEEN THE CHARACTERS

In the great tradition of the first Baldur's Gate, but also of the great RPGs of a now bygone era, such as Knight of the Old Republic, Dragon Age, Mass Effect and others, it is possible to bud a relationship between your character main and party members, this can be strictly friendly, or go further in Baldur's Gate 3. It's not easy, but it's possible to start these romances with all the companions on early access if you don't You are not the victim of a nasty bug.

A more occasional carnal relationship can also be initiated with a character outside the group on occasion. Let us recall in passing that Baldur's Gate 3 is a PEGI 18+ game, and that it has been openly confirmed that sexual content is present in the game. By launching the game, then in the options, you can choose to activate or not nudity. If enabled, expect to see sausages occasionally.

At the moment, all the characters seem to be Playersexuals, meaning they are willing to start a relationship with the player, regardless of race, gender, and class, as unlikely as that be able to show up.

When is romance possible?

Before you decide to look for true love, know that there are a few main rules that will help you determine whether romance is possible or not:

- A romance between the protagonist you play and an NPC is possible. Whether it's a custom character or an original character.

- A romance between two NPCs is not possible. Lae'Zel and Halsin aren't going to make out while you're busy elsewhere. It is possible to initiate elements of this type by directly taking control of these characters during dialogue, but the repercussions are generally quite strange and unpredictable.

- A romance between two player-controlled characters in cooperative mode is not possible. As the studio points out, nothing stops said players from kissing each other in front of the screen.

- It is possible to have multiple romances at the same time, but you have to expect negative repercussions or consequences at some point. It depends quite heavily on the companion you are having a romance with, some are open to multiple relationships, even threesomes. Others, not at all.

Recruit companions and confirmed romances

The first step is obviously to recruit said companions, or at least the one with whom you wish to start a relationship. They are all recruited from the large open world map of Act 1.

- Shadowheart is to the right of the beach after the ship crash. She can be difficult to recruit if you are playing a Gith. It can also be released from its capsule aboard the Nautiloid from a patch.

- Astarion is on the path along the cliffs to the left after passing the first group of intellect devourers. Agree to help him and he will accompany you.

- Lae'zel is in a cage guarded by 2 Tieflings to the right and above the dying Mind Flayer. You have to convince the Tieflings to free her, or defeat them with her.

- Gayle appears on the path that leads to the top of the cliff overlooking the beach. You have to help him get out of the rock.

- Wyll is a little more complicated to recruit since he must survive the big battle with the goblins in front of the Druid Grove. Exposing your characters and playing aggressively can help keep him alive. Then talk to him in the grove.

- **Karlach**: Wyll will ask you to kill her, which complicates her recruitment. It is located by the river, on the other side of the map. She agrees to join you if you kill the fake Paladins of Tyr nearby. You must then force Wyll to accept it, or kill him.

- **Halsin**: The druid leader is a prisoner of the goblin fortress. You must free him and kill the goblin leaders of the Absolute so that he agrees to join you. This makes recruiting Minthara impossible. You must lift the curse during chapter 2 to make it a permanent companion.

- **Minthara**: A paladin of Loth, present in the goblin camp. She agrees to join you if you massacre the druid camp and then save her in the prison of chapter 2. You can also recruit her by completely ignoring the grove and the goblins. Suffice to say that this makes the recruitment of Halsin impossible.

- **Minsk and Jaheira** : These old companions from the license are recruited later during the adventure. A romance with one of them is not possible.

Note: It is possible to have intimate relationships with other characters during the story, such as Mizora, Emperor, an Incubus or a duo of Dark Elves at Sharess's Caress, but we cannot really talk about a romance.

Approval of Companions

Once a companion is recruited, he will judge your actions and your words, whether or not he is in the active group. So you can't send someone back to camp to do your nasty little business behind their back. We might as well say it right away, it is difficult to satisfy everyone unless you know which option to choose, or play on the composition of the group each time. BG3's companions are quite temperamental, and some even have an openly conflicted relationship like Shadowheart and Lae'zel Recruiting one will anger the other, satisfying one will displease the other, etc. The famous "Shadowheart approves" and "Lae'zel disapproves" which appear at the top of the screen. Fortunately, it gets better afterward.

Almost everyone approves when you pet the dog. - Baldur's Gate 3

Almost everyone approves when you pet the dog.

To be able to start a relationship with one of your companions, their approval rating must be high enough. There are also certain dialogues and key answers that can permanently close this option. So be careful with what you say and keep in mind the character of the companion you want to seduce. If in doubt, save the game, it is possible even during a dialogue. Load the game if the result does not suit you.

You can see what active party members think of you in the characters tab.

Progression in history

As is often the case, progress in the story as well as rest at camp serve to determine when romance is available. It is advisable to rest often in order to be able to participate in the various dialogues and events linked to each companion, which will also allow you to improve your reputation if necessary.

However, the decisive element is the quest to free the druid Halsin in the goblin fortress (door at the back right then kill the leaders). By knowing where to go and what to do, it is possible to reach it quickly, but normally it requires 20 to 25 hours of gameplay. After the festivities in the Druid Grove, you can talk to your companions and suggest they spend some time together, or choose to sleep alone. If they like you enough, they should accept your offer.

We'll list ways to improve approval for each companion on the following pages. As well as what causes a loss of approval. This is not an exhaustive list. Don't forget to save regularly (F5) in order to load the game if the result is not satisfactory. Actions that require particular skill rolls are by definition random.

We don't know at this time if it will be possible to have multiple romances going on in parallel, but given that the camp characters seem to be spying on each other, it seems unlikely. Your dream of seducing both Shadowheart and Lae'zel will quickly turn into a nightmare.

Romance with Shadowheart

This priestess of Shar tries to hide the fact that she worships this evil goddess, as well as many other secrets. She doesn't like people trying to find out about them. She doesn't like the Githankis and Lae'zel in particular, generally pleasing one will anger the other. Although she appears to be of rather evil alignment, she prefers subtlety and finesse, as well as avoiding unnecessary fights. She also appreciates that you show determination and strength of character. By succeeding in a few fairly difficult Persuasion rolls, or with the right clues and events, you can discover his faith. Accepting it and then talking about it in a positive way makes it easy to get closer to her. She does not like gratuitous cruelty, weakness and stupidity.

Attention : During chapter 2, you have to take Shadowheart to Shar's Gauntlet, and you have to convince her not to kill Chantenuit one way or the other. If she stays faithful to Shar, your romance will end badly.

❖ *Shadowheart approves*

- Tell him that finding a caregiver is the priority.
- Show Guex your sword technique in the Druid Enclave.
- Persuade Arka that killing the caged goblin Sazza will make her take her secrets to the grave.
- Then speak to Sazza, when she mentions the Gut healer, speak to Shadowheart and tell him that you must study all avenues, even the most repulsive ones.
- Rescue Nadira from the Hobgoblin assassin in the Druid camp, then trick her into giving you the coin.
- Promise Nettie to take the poison.
- Resist the Absolute during the first dream.
- Refuse Raphael the Devil's offer. When you next speak to Shadowheart, always be firm in your refusal of his offer.
- Tell Andrick and Brynna to forget Owlbear and leave.
- Crush the tadpole that comes out of the body next.
- Convince Owlbear to let you go.
- Tell Scratch the Dog that he can come to your camp if his owner doesn't wake up (using animal language).
- Face the false god in the Fishman Cave.
- Throw shit at the goblins at the entrance to the ruins.
- Persuade the goblin to kiss your feet in the goblin camp.
- Petting the dog at camp.

❖ Shadowheart disapproves

● Recruit Lae'zel.

● Go in Lae'zel's direction during dialogues.

● Telling her she's hiding something during a conversation at camp.

● Agree to help Wyll save the refugees.

● Open the barn in which a hobgoblin and an ogress are fornicating in the village

● Tell Auntie Ethel you have a tadpole in your head.

● Helping Ethel when Johl and Demir confront her about their sister in the woods.

● Try to persuade the goblin guard in front of the camp.

● Smear shit on your face when Olak the Goblin asks you to.

● Condemn his faith in Shar.

Romance with Astarion

Astarion is a high elf spawn of a vampire, discovering this specific point and accepting it, or even donating your blood will seriously accelerate the romance. It should be noted that Astarion is certainly the most openly evil and cruel character of the group. Degenerate, cruel acts, lies, even self-destructive behavior tend to get his approval. He's certainly the most isolated character of the group in this area, and it's going to be difficult to gain his approval at the same time as that of the others.

Attention : During chapter 2, mishandling Araj's quest will push Astarion to leave you, unless you properly manage the dialogue that follows. You must also manage your choices carefully during the confrontation with Cazador, his vampire master, during chapter 3, and the ritual that follows. It is better to convince Astarion not to perform the ritual.

❖ Astarion approves

● Trick the Tieflings into releasing Lae'zel, then have her plead with you before releasing her. Or join her to slaughter the tiefelins.

● Explain your tadpole problem to Auntie Ethel in the Druid Grove.

● Let Arka kill Sazza the caged goblin in the druid camp.

● Then tell Astarion that the refugees are desperate and ready for anything next.

● Intimidate or provoke human and Zevlor adventurers in the Druid Grove.

● Kick the squirrel into the Druid Grove.

● Tell Kagha the Archdruid that you appreciated her demonstration when she asks you if you think she is a monster.

- Sting to death the bird Nettie is treating.

- Attack the apprentice Druidess Nettie after the poisoning.

- Agree to kill Kagha without trying to persuade her.

- Try to persuade the guards at the entrance to the abandoned village.

- Give him Thay's book of necromancy in the abandoned village laboratory.

- Open the door to the barn in which a Hobgoblin and an Ogress are fornicating in the abandoned village.

- Ask for a reward from the gnome attached to the windmill after saving him.

- Crush the tadpole that comes out of Edowin's head in the woods. Then tell Astarion that you're practically worshiped and that might be helpful.

- Refuse Raphael the Devil's invitation to remove your tadpole.

- Tell Astarion that you want to die like him, beheaded with an axe.

- Let Astarion drink your blood (be careful, risk of death).

- Support Astarion the next morning, after his vampire nature was revealed, during the dialogue with the group.

- Tell him he can drink the blood of your enemies.

- Persuade him to share his dream with you.

- After the altercation between Mairina and Ethel's brothers, tell them that they have to manage.

- Kill Gandrel, the monster hunter next to Ethel's hut in the swamp.

- When talking to Lorin the elf in Ethel's cave, pretend to be the monster he sees.

- Spare Ethel when she surrenders.

- Bring Mairina's husband back to life, give him the staff and tell Astarion it's fun.

- Support Astarion's decision to continue using the tadpole's power.

- Don't apologize after using the power again during dialogue in the camp.

- Ignore Volo in the goblin camp.

- Feed the Owl Bear.

- Kiss the goblin's feet in the camp and steal the ring.

- Volunteer to torture Liam in the Goblin Fortress.

- Tell Abdirak that you agreed to be tortured in the goblin fortress.

- Accept Minthara's offer to attack the Druid Grove.

- Agree to open the door to the grove for the goblins, and do so.

- Confront the False God in the Fishman Zone of the Shadowlands, or become his Chosen and be worshiped.

- Kill the Githyanki Patrol (Difficult Fight)

- Agree to share your bed with him after the Tieflings party.

- Choose the option of presenting your neck to him to invite him into bed.

- Agree to kill duergar for the Myconid leader in Shadowlands.

- In the Zentharim Smugglers' Cave, buy the artist and tell him he is your slave.

❖ Astarion disapproves

- Agree to help Wyll save the refugees.

- Questioning his attitude towards the mind-controlled fishermen in the Nautiloid wreck.

- Saying the Druids went too far in talking to Marriko and Locke in the Druid Enclave.

- Agree to help Zevlor.

- Take the poison from Nettie the Druidess.

- Promise Nettie that you will take the poison if you feel any symptoms.

- Agree to help Marina and Morina.

- Read Thay's Tome of Necromancy for yourself in the village.

- Persuade Zevlor and the human adventurer to calm down.

- Reveal Astarion's identity to the monster hunter near Ethel's house.

- Fail the persuasion roll to make him reveal his dream.

- Kill Mayrina's zombie husband after raising him with Ethel's wand.

- You smear shit on your face in the goblin camp when Olak asks you to.

- Let Gut cleanse your mind in the Goblin Fortress.

Romance with Gayle

Although he is quite pretentious, even narcissistic, Gayle is probably one of the companions whose alignment seems to be closest to "good". He appreciates it when you help others (and him in particular). You don't need to act like a paladin either, but saving people you come across and giving him artifacts to calm the magical time bomb in his chest is a great way to improve your reputation . Avoid delving into his thoughts with your tadpole on the other hand, he hates it. Gale also has a very funny special quest that begins when he dies to bring him back to life by following the instructions. Following them and resurrecting him is another way to gain a lot of approval from him, as well as push him to open up to you.

❖ Gayle approves

- Seek a cure as a priority during camp dialogues.
- Convince the Tiefelin Rolan to stay in the Druid Enclave.
- Agree to help Wyll save the refugees.
- Practice your sword techniques in Guex in the Druid Enclave.
- Saying the Druids went too far in talking to Marrijko and Locke in the Druid Enclave.
- Refuse Raphael the Devil's offer.
- Crush the tadpole that comes out of the corpse.
- Give him Thay's Tome of Necromancy.
- Give him magic items to absorb.
- Bring him back to life by following the procedure.
- Convince the Owl Bear not to kill you in his cave.
- Convince Owlbear to let you go.
- Tell Scratch the Dog that he can come to your camp if his owner doesn't wake up (using animal language).
- Help Auntie Ethel when Johl and Demir confront her in the forest.
- Agree to help Zevlor.
- Persuade the goblin to kiss your feet in the goblin camp.
- Help Karlach.
- Fight the false god in the fishman cave.

❖ Gayle disapproves

- Help Lae'zel kill the Tieflings after persuading them to release her.
- Read Thay's Tome of Necromancy for yourself.
- You smear shit on your face when Olak asks you to.
- Open the door to the druid enclave at the request of the goblins.
- Behave cruelly.
- Reading thoughts with the tadpole.

Romance with Lae'zel

It is not easy to make yourself liked by Lae'zel unless you adopt a particular approach to the game. You generally have to go in his direction and not hesitate to show outright violence or at least firmness. You must also demonstrate strength of character during various trials and dilemmas. As noted on a previous page, she has a conflicted relationship with Shadowheart, the two get along very poorly and what offers approval for one will earn you disapproval for the other in many cases.

Attention : You have to manage the different dialogues in the Githyanki Crèche carefully. Show respect to the queen, but also push Lae'Zel to question her teachings in order to push her to rebel later. You will also need to free Prince Orphéys at the end of the game, if you want a happy conclusion with her.

❖ Lae'zel approved

● At camp, tell Lae'zel that you will find a cure.

● Choose to directly attack Taman and Gimblelock during the dialogue at the entrance to the cliff crypt.

● Tell Zevlor you don't have time to help him in the Druid Enclave.

● Show off your sword fighting skills in Guex in the Druid Enclave.

● Force the Tiefelin to obey Lae'zel when she questions him.

● Tell Kagha that she only protected her people after the snake killed the little girl.

● Promise Nettie to take the poison.

● Refuse Raphael the Devil's invitation to remove your tadpole.

● Resist during the dream of the Absolute.

● Refuse to use the tadpole's powers any further.

● Ignore Volo after talking to him in the Goblin Camp.

● Force the goblin to kiss your feet.

● Fight the false god in the fishman cave.

● Let her talk to the Githyanki patrol.

❖ Lae'zel disapproves

● Convince the owlbear not to kill you.

● Succeed in convincing Kagha not to kill the child.

● Try to persuade the goblin guards at the camp entrance.

● Smear shit on your face at Olak's request.

● Agree with Shadowheart or listen to his advice.

● Behaving kindly or weakly.

Romance with Wyll

Wyll is the other rather good character in the group, despite his troubled past. He likes to help people in need, however he has a very antagonistic relationship with the goblins who will not fail to recognize him directly. If you want to get into Wyll's good graces, allying with the goblins, or negotiating with them, will rarely be an option. Other than that, he's pretty tolerant one way or the other, as long as you don't slaughter Tieflings and aren't too friendly with Mizora. You will also have to manage the rescue of the Grand Duke well.

❖ Will approve

● Show off your sword skills in Guex in the Druid Enclave.

● Agree to help Zevlor.

● Refuse Raphael the Devil's offer.

● Crush the tadpole that comes out of the corpse.

● Tell Scratch the Dog that he can come to your camp if his owner doesn't wake up (using animal language).

● Persuade the goblin guards at the camp entrance.

● Agree to confront Fezzerk in the abandoned village.

● Convince the Owl Bear not to kill you in his cave.

● Convince Owlbear to let you go.

● Persuade the goblin to kiss your feet in the goblin camp.

● Help Karlach.

● Petting the dog at camp.

❖ Wyll disapproves

● Open the barn door in which a hobgoblin and an ogress are fornicating.

● Smear shit on your face at Olak's request.

● Open the door to the druid enclave for the goblins.

● He will leave the group if you join the goblin faction.

● Behaving cruelly in general.

Romance with Karlach

The last original character and companion introduced into the game is Karlach, a very muscular tiefling as well. She's a Barbarian with an infernal machine for a heart, which makes her very, very hot (literally, since she catches fire). Details about the character's history and romance are still poorly known, but from what we have seen, she has a very frank and open personality, and she is of good alignment. Situations must be resolved through diplomacy, when possible, rather than through deception. She has nothing against the use of force, but not to kill innocent people. You should not massacre the Tieflings in the Druid Grove, which seems rather obvious. You also have to try to solve your heart problems (literally for once) as well as overheating.

Romance with Minthara

As mentioned earlier, not all relationships involve a mate. In this case "romance" is not the right word to begin with, since it is simply a one-night stand with the dark elf priestess Minthara who leads the goblins in the fortress. The operation is radically different since it is an NPC. You have to infiltrate the fortress in one way or another and agree to help kill the Tieflings and Druids. This opens a new questline that involves decimating all the inhabitants of the grove using the goblins Be warned, it's quite long and tedious as is, as well as a little buggy on occasion.

Once the grove is destroyed and the celebration begins, Minthara will suggest that you celebrate in a more intimate way (even if you play a dwarf). During the scene that follows, you can choose to search his thoughts. It is better to look for her desires so that you can persuade her in the conversation that follows. Minthara does not immediately become a member of your party. She must be rescued during chapter 2, in Hautelune prison. It must then be freed from the influence of the Absolute. It then becomes possible to start a more serious relationship.

Minthara is ruthless and cruel, it is easier to gain her approval by playing an evil character, like Dark Urges.

Attention : While playing Dark Impulses, rejecting your inheritance in the Temple of Bhaal will cause her to leave you.

Romance with Halsin

The leader of the grove's druids is an incredibly strong Wood Elf named Halsin. He is a shapeshifting druid, who loves transforming into a bear to kill his enemies. For the moment, relatively little information is available on his romance, but we can assume that respecting nature and protecting it are good ways to attract his favors. It was confirmed during a Larian Studios stream that a romance is possible with him, including playing Astarion. There are even extremely exotic options, since it's possible to ask him to transform into a bear during the act, if that's the kind of experience you're looking for.

You must correctly deal with the curse in Chapter 2, with Art Cullagh, Thaniel and Oliver, in order to be able to recruit Halsin. Once in Chapter 3, starting a relationship is then easy, as long as your approval gauge is high enough. He is also open to fairly naughty adventures with some of your companions, if you are already in a romance with Shadowheart for example.

SHADOWHEART

Even if she seems to bring together all the clichés of gothic emo at the beginning of the story with her pseudonym and her love of darkness, Shadowheart is a character who can greatly evolve and reveal another nature in Baldur's Gate 3.

How to free Shadowheart from the Illithid crate?

During the tutorial, you will come across Shadowheart while she is still a prisoner. It is strongly recommended to free her in order to kill Commander Zhalk, and also gain his recognition. You can find the required rune in the next room, on the dead slave. All you have to do is use the larva's powers on the console.

How to recruit Shadowheart?

After the end of the tutorial, you will regain consciousness on the beach which marks the start of chapter 1. If you have freed it, Shadowheart will be unconscious a few meters from you. Waking her up will allow you to recruit her. Otherwise, it is a little further away, in front of a locked door.

What should be done with the relic?

After Shadowheart is freed, you can activate the relic in his inventory, which triggers a dialogue that seemingly leads nowhere. You can also attempt to take the relic when Shadowheart is unconscious.

There is no point forcing Shadowheart to give up the relic, or talk about it at this time. It is closely linked to the main story, and you have to wait for special events for the situation to evolve. You also don't risk losing it, or having it stolen. But showing it to those you interact with (goblins, Githyanki, etc.) has a strong tendency to provoke a fight.

Shadowheart Utility

Shadowheart has a great class, but its stats and divine domain are just horrible. It is advisable to find Blight and pay Shadowheart a respec, at least, if it does not go against your beliefs. There are many possibilities, depending on the role you want her to play, whether that of a fantastic healer with the life domain, or a spellcaster with the light domain. It can also be more formidable in melee with the domain of war. You can increase his strength, while reducing his dexterity and charisma in order to make him wear heavy armor, in order to make him a good tank. As a cleric Shadowheart proves useful in combat with his healing, buffs, summons, and offensive spells.

Outside of combat, Shadowheart shines with his Assist spell, his various utility spells, and his high Wisdom score giving him Perception bonuses.

Relationship with Shadowheart

It is advisable to keep Shadowheart in the active group to accelerate reputation gains. You can charge the party and send her back to camp if you are going to perform an action she disapproves of. At the start of the game, Shadowheart really hides a lot of things from you. You will need to penetrate her shell, and succeed in a few persuasion rolls before you really start to get to know her and accumulate the relationship bonuses. She will end up confessing to you that she is a priestess of Shar, the evil goddess of darkness, so it becomes easier to speak to her with an open (shadow) heart.

Shadowheart approves

- When you respect her secrets, never insist too much when she is hiding something from you.

- That you respect her goddess and her religious beliefs.

- Intelligent, clever and discreet decisions and actions to solve problems, rather than brute force.

- Strangely, she appreciates good deeds in general. She likes it when you help people, especially if you think about demanding a reward afterwards.

- At the start of the game, she approves when Lae'Zel disapproves.

Shadowheart disapproves

- When you insist on knowing his secrets.

- If you insult her goddess or her faith.

- When you defend Lae'Zel, or perform an action she approves of.

- If you do stupid, degrading or self-destructive actions.

Romance

Since there are many opportunities to improve your relationship with Shadowheart it is quite easy to start a romance with her. But it also risks derailing along the way and failing if you don't make the right choices. As mentioned above, she should reveal to you that she is a priestess of Shar, and you shouldn't make a big deal out of it. When you talk to her at camp, you'll end up being given additional options to invite her to talk more about herself, since you haven't really been able to truly connect with each other.

- You've truly begun your romance with Shadowheart when she invites you for a drink on the cliff at night. Try to understand his intentions and then kiss him.

- During the celebration party, after the defeat of the goblins (if applicable), consider going to talk to him and make advances towards him. Then go to sleep and dream about her. Unfortunately for horny players, the erotic scene is one of the last to arrive in the game, if not the last.

- We will detail a little further, but chapter 2 is an important turning point in your relationship, you absolutely must have Shadowheart in the group when you perform the Gauntlet of Shar, and especially, when you face Balthazar to reach Nightsong. We must convince Shadowheart not to kill Chantenuit. If she kills her, everything indicates that the romance falls through, since she will devote herself to Shar. You will need to support Shadowheart during his crisis of faith, and the transition that follows at the end of chapter 2 and the beginning of chapter 3.

- During chapter 3, Shadowheart will change his hairstyle and attitude. And she will end up offering you a night meeting, but only if you camp at the entrance to Baldur's Gate, in the wild. If you unlocked the room at the inn, this won't happen, which can be disappointing. Consider changing zones and increasing the number of long rests.

- From this point on, everything should be fine. If Shadowheart is the first person you spend the night with, she will even offer to spice things up with other characters like Halsin. She will also fully accept the fact of having spent the night with Mizora for example.

Gauntlet de Shar

This massive section of chapter 2 is best done with Shadowheart. She will take the initiative regularly, with special dialogues. She will also obtain the Spear of Night in the Library of Silence.

As mentioned above, the decisive turning point is facing Chantenuit. We must not let Balthazar take it away. If you decide to encourage Shadowheart to kill Nightsong, she will then devote herself to the goddess Shar. In addition to blocking his romance, this will cause him to take a darker path. If you persuade her to spare Chantenuit, she then begins her redemption, with quite a few changes to come during the next chapter.

You will then be treated to several dialogues involving Chantenuit and Shadowheart. She will also receive a new legendary spear, regardless of her choice.

House of Sorrow: Should Shadowheart be delivered to Viconia?

Upon entering Baldur's Gate, Shadowheart will set out in search of the temple of Shar present in the city. It is located in the northwest corner of the lower town, under the guise of a care home. These institutions definitely have serious problems in this game. The reasons for going there are obviously going to be diametrically opposed, depending on your decisions regarding Chantenuit.

- The critical moment is the meeting with Viconia. She will ask you to deliver Shadowheart to her, which you can do, in order to obtain the support of the Sharists against the Absolute. If you refuse, a big fight begins. Needless to say, supporting the Sharists is clearly not worth the sacrifice of one of your companions?

- If Shadowheart is still loyal to Shar, and she has killed Nightsong, then she may choose to overthrow Viconia as leader of the cult. Some of the cultists present will then assist you during the fight.

After the battle, you can search the area, which will allow Shadowheart to reconnect with his past. Let it be in the form of a cave containing the flowers she prefers. There is also her former best friend, whom she has forgotten. But the important point is the presence of his parents, behind the big door of the temple.

Should Shadowheart's parents be killed?

If Shadowheart is still loyal to Shar, the question doesn't really arise. She can perform them without batting an eyelid. Conversely, if she found redemption, the question is legitimate. The "right" choice is probably to spare his parents. Thus, a more positive conclusion will be available at the end of the game.

A notable bug encountered by many players is that even after killing Shadowheart's parents, the curse on his hand did not disappear. It's no luck !

Conclusion

As long as you don't choose the evil ending of Baldur's Gate 3, the endings available for Shadowheart are relatively simple. They were not presented before the deployment of a recent patch, it is possible that you missed them by finishing the game previously:

● Shadowheart can remain loyal to Shar and serve her at the head of the cult, as Viconia did.

● If her parents are alive, she will build them a house in a quiet place. In case of romance with Shadowheart, she invites you to live with her.

ASTARION

Absolutely no one is normal in your group in Baldur's Gate 3, except you obviously! At least, if you don't play Dark Impulses, then you're the worst of the lot. But let's get back to the charming and cruel vampire.

How to recruit Astarion?

Unless you play it yourself, Astarion will quickly join your party at the start of Chapter 1. Just take a detour to the west of the beach and the nautiloid wreck. A secret, the Chipped Stone is also nearby. The first contact is not very good, as he tries to take you hostage and stab you. It's better to forget this slight detail and accept him into your group, everything should be fine?

Usefulness of Astarion in the group

As a Dodger, Astarion is an almost obligatory member of the party. Unless you play a character with similar skills (Rogue, Bard or Ranger), you may have difficulty managing traps, locked locks, theft and infiltration phases. Astarion can handle all of this, as well as Perception, and even dialogue to some extent.

If played well, Astarion is formidable in combat, thanks to his sneak attacks. It's easiest to play him with a bow, have him stealth, and then deal a devastating attack to an enemy. This also works in melee, if the target is the victim of a control effect, one of your companions is nearby, or Astarion is camouflaged. Choosing to make Astarion an Assassin is probably the best choice to optimize him in combat. You can multiclass him into Warrior or Ranger, to give him a second attack and some additional bonuses.

Revealing Astarion's identity

You probably already know that Astarion is a vampire, but your character isn't. Before you can truly begin to progress in your relationship with Astarion, you must force him to reveal his identity. The best way to do this is to inspect the dead animals in the area, especially the wild boar at the entrance to the ruined village in the center of the map. Then use a long rest. When you wake up, you will find Astarion about to suck your blood.

The "right" choice is not to wake the whole camp, and let them drink your blood. However, you will have to manage to interrupt him in time, it is possible that he will kill you this way. It is therefore better to make a backup beforehand. Worst case scenario, control another companion to bring you back to life with the help of Wither.

A discussion should take place upon awakening (if you have not died miserably) in order to reveal Astarion's identity, and to decide his fate. It is better to defend him, and encourage him to drink the blood of your enemies. Limiting it to animal blood can have negative repercussions on Scratchy and Owl Bear.

Improve your relationship with Astarion

Like your other companions, Astarion has his own values. We might as well tell you clearly, it can be considered to be of bad alignment. He enjoys cruelty and evil acts, especially if they are funny. Tormenting, humiliating, torturing or torturing someone is an activity they will enjoy. Allying with goblins or monsters is a good way to gain favor. In fact, it is rather difficult to quickly gain approval by playing a good character. Voluntarily offering your blood when he asks for it is a good way to quickly progress in his heart.

He is not completely opposed to kind acts, but not to the point of turning you into a victim or a selfless hero. He also approves of loyalty to his person, and that we ourselves approve of his words. Standing up for him whenever possible, and saying that he was right to act in one way or another is a good alternative to overtly malicious actions.

The good thing about Astarion is that he doesn't disapprove of much in general. You should therefore gradually climb in his esteem. Also remember to help him when he tries to see the scars on his back.

Romance

Astarion is one of the easiest and quickest romances to unlock, as long as your relationship is high enough during Chapter 1. Follow the instructions in the previous paragraph to achieve this. He will end up talking to you at camp and making advances towards you. As long as you don't reject him, your relationship should continue to grow. He will offer you different evening meetings, in which you can participate.

Astarion's Unique Quests

In addition to his original quest with Cazador, which we'll see later, Astarion is tied to two minor quests. The first takes place during Chapter 1, with Gandrel, a vampire hunter located next to Auntie Ethel's house. You must let Astarion kill him or initiate combat on your own initiative, in order to obtain his approval. Delivering Astarion or denouncing him will cause him to leave your party permanently.

The second quest can be found on the ground floor of the Hautelune Towers. Araj Oblodra, a rather shady drow, admits to you that her fantasy is that a vampire drinks her blood. She asks you to return with Astarion to your group, if he is not already present. She offers you a potion giving a permanent +2 Strength to a character, if Astarion agrees to drink her blood. You will have to convince him to do it, which he disapproves of. But it's a small price to pay.

Let's spawn vampires

You must reach Baldur's Gate in chapter 3 before really progressing in Astarion's personal quest. While visiting the inn located on the bridge leading to the city, you will come across two other offspring of Cazador. How you handle the conversation doesn't matter too much, since they're bound to succeed in escaping. Don't be afraid to encourage Astarion to be cruel and ruthless in order to gain approval.

Cazador and the ritual

During chapter 3, it is advisable to attack Cazador Castle in the company of Astarion. To reach the entrance, you must pass through the guard tower, in the central part of the lower town. Then, follow the ramparts towards the north, to reach the back door.

We are not going to detail here this rather long and complex passage of the game. You can discover it in the guide below. But in summary, three outcomes are possible for Astarion:

● Astarion kills Cazador, but he does not use the blood ritual. It is the path to redemption in order to become someone better, and less cruel.

● If you begin the ritual, but then interrupt it, you will have to kill Astarion, without the possibility of bringing him back to life.

● Completing the ritual gives Astarion a large power bonus, as well as a rather different conclusion. But his eyes now glow with a disturbing red glow, he also becomes more malevolent.

End of the game and conclusion

If he is alive at the end of the story, Astarion has two main endings, which depend greatly on whether or not you participate in Cazador's ritual. If you haven't gone to see the master vampire, or you haven't completed the ritual. Shortly after the death of the Absolute, Astarion begins to burn in the sun again, which forces him to flee, taking refuge in the darkness.

If Astarion has completed the ritual, he can stay in the sun and participate in the festivities, but his much more evil nature than before does not bode well.

LAE'ZEL

In addition to being highlighted in the introductory cinematic of Baldur's Gate 3, and having a very strong character, Lae'Zel is also a real killing machine. It is therefore not surprising that she has both fans and detractors.

How to recruit Lae'Zel?

If you don't play as the original character, Lae'Zel is literally your first companion aboard the Nautiloid. But you lose track of him after confronting Commander Zhalk, and crashing on the beach. Fortunately, it doesn't take long for you to be reunited. She is trapped in a cage, with two Tieflings wanting to kill her. Be careful, this is a critical passage: by default, you must choose between killing the Tieffelin or Lae'Zel. It is better to try to persuade them to flee with skill rolls. Killing them can cause you to break your Paladin Oath.

Usefulness of Lae'Zel

As a Warrior, Lae'Zel serves as both a tank and physical DPS. With a good two-handed sword and a few levels, then the Speed spell, she transforms into a combine harvester mowing down enemies like wheat. His presence is required for several important quests in the game, and there are many items offering bonuses exclusive to the member of his race.

Outside of combat, his considerable strength also allows him to serve as a mule, move heavy objects like the Chipped Rock, and jump long distances. Her Githyanki racial bonuses amplify this further. His ability to arcane knowledge also helps him master skills that you may need, and that your group lacks, for example, those based on Dexterity if you don't have a Rogue (poor Astarion), or those based on Intelligence if there is no Gayle.

Relationship with Lae'Zel

At first glance, Lae'Zel seems very difficult to please, and of evil alignment, but it's a little more complicated than that. It is advisable to have Lae'Zel in the group the majority of the time, in order to have the opportunity to gain his approval. Send her back to camp when you know your choices are going to be disapproved.

❖ Lae'Zel approved

- She enjoys dominating and humiliating her opponents, as well as commanding in general. Solving your problems through violence is by far her favorite.

- Lae'Zel likes to focus on the mission and your goals. Keep this in mind during dialogue.

- At the start of the game, his relationship with Shadowheart is very bad. She approves when Shadowheart disapproves.

- She considers the Githyanki to be superior in every way. Going your way is always received positively in this area.

- Following militaristic rules is a good way to gain approval. You must respect the hierarchy, follow orders, and not hesitate to sacrifice yourself.

- Let Lae'Zel handle all Githyanki-related situations (patrol, nursery, Voss).

❖ Lae'Zel disapproves

- She hates cowardice and weakness.

- At the start of the game, she disapproves when Shadowheart approves.

- Insisting on speaking for him in situations involving Githyanki.

- There are many situations in which Lae'Zel should disapprove, or even become downright hostile, but you can keep the situation under control by explaining your motivations to him. Succeeding at a persuasion roll is also a good way to avoid an undesirable outcome.

Romance

Ironically, Lae'Zel's romance is the one that begins most quickly, and most directly, with sex. We can say that she is direct. It's not for nothing that there is a speedrun in this category. By gaining a little approval, she will very openly make advances towards you, which you must accept, otherwise you will permanently block her romance. However, she will then tell you that it was just a one-time thing, and that no feelings were involved.

Things get a little complicated then, during chapter 2, she will challenge you to a duel during a long rest. It may be wise to deprive her of her equipment before resting, if you really want to win the fight. She is tough after all. In case of victory or defeat, you must respect her wish and say that she is yours, or that you are hers. Things then take a more romantic turn and she will even kiss you on request.

Just try not to betray her cause, or deceive her. Some companions accept the fact that you are unfaithful, but this is absolutely not the case with Lae'Zel. She may reject you directly between chapter 1 and 2, or during chapter 3, if you sleep with the Dark Elves or Mizora for example.

Githyanki Patrol

In the northwest corner of the Chapter 1 map, you will encounter a Githyanki patrol with a red dragon. It's better to have Lae'Zel in the group, and let him lead the negotiations. You can encourage him to deceive his peers, but if you are not afraid of a little disapproval, it is also possible to initiate hostilities by showing the relic for example. Be prepared to have to persuade Lae'Zel either way. Taking down the Githyanki Patrol is a great source of gear and experience.

Githyanki Nativity Scene

As you enter the mountain pass with Lae'Zel, she will guide you to the Githyanki Nativity Scene. Once again, let her lead the negotiations. You will have many opportunities to gain approval with Lae'Zel in the zone, or lose it:

● Letting her get on the larva extraction machine first is appreciated, it also makes her easier to persuade to betray Vlaakith later.

● It's best to kill the Inquisitor and his group directly when you encounter them, this limits the chances of Lae'Zel turning against you.

● Show respect to Vlaakith, and pretend to accept his orders. Then enter the portal to the Astral Plane.

● After the meeting with the nocturnal visitor, you will have to convince Lae'Zel that Vlaakith has betrayed his people, and that they must rebel. Of course, you can also encourage him to remain loyal to his queen, but this leads to much less favorable and, above all, less profitable events.

● You will have to kill the occupants of the nursery when leaving the Astral plane do not hesitate to massacre everyone.

Rebellion against Vlaakith

During the following long rest, Kith'Rak Voss and his companions will invite themselves to the camp, to negotiate with Lae'Zel. Once again, it is best to encourage him to rebel against Vlaakith. At least, for a more favorable outcome.

The identity of the Emperor and Prince Orpheys

During the passage to chapter 3, a nocturnal event asks you to borrow a portal created by the nocturnal visitor, in order to help her. There is a group of Githyanki there, as well as a Mind Flayer. Help the Emperor kill the invaders, then listen to his explanations, there is nothing more to do at the moment.

Chapter 3: Raphael and the Orphic hammer

You will find Kith'Rak Voss in the caress of Sharess, in the middle of negotiations with Raphaël, our favorite fiend. He will offer to exchange the Crown of Karsus for an artifact capable of freeing the prince, a legendary object. If you refuse, Lae'Zel and Kith'Rak Voss will protest, but there is a better solution. Invite yourself into the House of Hope and steal the hammer yourself. Once you have the hammer, you can show it to Voss in the sewers. He will give you the Silver Sword of the Astral Plane, the ultimate weapon for Lae'Zel, as thanks.

Orin and Lae'Zel taken hostage

Be careful, if Lae'Zel is important to you, make sure to keep her in your group after Gortash's coronation in Chapter 3. In the event of a long rest, or upon reaching a certain area of   the sewers, an event will occur. produce, and one of your companions will be kidnapped by Orin. Lae'Zel is the default victim, if she is not in the group. She risks disappearing until the end of chapter 3.

Free Prince Orpheys

Once in possession of the hammer, it is legitimate to seek how to free it. This is not possible until the final battle, however. You can find the details in our guide to Prince Orphéys, and the associated choices, such as transforming him into a Mind Flayer or not. Suffice it to say that if you want to favor Lae'Zel, it is better to free the prince, and if possible, not transform him into a monster. Even if persuasion rolls allow him to accept other choices. Know, however, that she will inevitably leave you if you devour the prince, or allow him to be devoured.

Conclusion

Lae'Zel is one of the characters with the greatest number of variations depending on the choices made. Here are the main purposes, as far as it is concerned:

- She may have been killed by you or Orin during the story.

- If she remained faithful to Vlaakith, she returns to serve her queen.

- She can also lead the rebellion against Vlaakith, if the prince has become a Mind Flayer.

- If the prince is alive and you are a couple, you can convince her to stay with you.

WILD

Wyll is a character who changed a bit between the early access phase of Baldur's Gate 3, and the full game. With a new quest, and a new past. He is one of the team members whose personal story has the most links to the main story, and you will have good reasons to interact with him, even if he is not on your active team .

Where and how to recruit Wyll?

Like the other original characters in the game, Wyll is recruited very quickly if you don't play him. He is seen for the first time at the entrance to the Druid Grove, during a skirmish against a group of goblins. His death is unlikely, but it remains possible, so you must try to shorten the fight and keep him alive. Then enter the Enclave. You will see Wyll training children in combat. All you have to do is talk to him and agree to help him solve his quest, chasing the Devil, so that he can accompany you.

Usefulness of Wyll in the group

As a Fiend Occultist, Wyll is primarily a spellcaster, although you can specialize him for melee combat with a pact weapon later on. He can be formidable at a distance as well as in melee in this case. This is a good choice of character to replace Gayle for example. It is much less versatile than the latter, but it is also much less dependent on its spell slots.

Outside of combat, Wyll can mainly serve as the face of the group, managing dialogue requiring persuasion or intimidation rolls, thanks to his high Charisma score. But if it's something your protagonist already handles, then it won't add much.

Should we kill or spare Karlach?

The famous she-devil can be found in the northeast of the map, by the river, and it is none other than Karlach, another original character who can be recruited. The ensuing dialogue changes greatly depending on whether Wyll is in the group or not If Wyll is present, he wants to kill Karlach, but you can convince him that she is not a she-devil, but just a Tiefling. You can then go and eliminate the false paladins of Tyre, in the building a little further north. Of course, you can also choose to kill Karlach to complete the mission. If Wyll is not present, you can recruit Karlach, and the dialogue between them will take place at the camp, during the next long rest. Unless you're really trying to play someone evil, it's best to kill Karlach. She's an excellent fighter, a potential romance, and one of players' favorite characters. But your choice will inevitably have repercussions.

Karlach-related consequences

You will have to wait several long rests before the consequences of your actions become known. Wyll will notify you when they arrive. Mizora, the real devil with whom Wyll made a pact.

If you spared and recruited Karlach, Mizora will punish Wyll. His appearance will take a hellish turn, with protruding horns. This is cosmetic, and in no way blocks Wyll's subsequent quests.

Conversely, if you killed Karlach, Mizora will give you a very rare Infernal Robe, allowing you to cast the Fire Shield spell (level 4), giving +1 AC and resistance to fire damage. However, this is far from compensating for the loss of Karlach.

How to improve your relationship with Wyll?

It is best for him to be active in the group, so that you have maximum opportunities to gain his approval. Wyll can be considered a good-aligned character. It's not for nothing that he has the title of Blade of the Borders, he is a local hero.

❖ Will approve

- Gaining his approval (or disapproval) is very simple and straightforward:
- Perform good deeds, even heroic deeds. Saving or helping others is always appreciated.
- Supporting him and helping him progress in his personal quest is appreciated.

❖ Wyll disapproves

● Selfish, evil, cruel or self-destructive actions.

● Consider making a pact with Raphaël (or even do it)

● Compliment Mizora, advocate for her, even spend the night with her. If you have a romance with Wyll, he will take it very badly.

● Ally with Minthara to destroy the Druid Grove. This will cause Wyll's permanent departure.

Romance with Wyll

Starting a romance with Wyll isn't very difficult, as long as your approval level is high enough. You must have solved his first personal quest with Karlach, one way or another. He will tell you about his origins, his life in high society, the fact that his father is the Grand Duke who has just been kidnapped, etc. He's also going to invite you to dance one night. It's an event that tends to occur to everyone, even those who haven't planned on starting a romance with Wyll. If you want a romantic relationship, you have to agree. For others, simply refusing will put an end to his advances.

It should be emphasized that several events can put a definitive end to your romance, such as allying with the goblins or killing Mizora during chapter 2.

Chapter 2: Saving the she-devil from Zariel

Once the second chapter of the story begins, Mizora will visit you a second time, and ask you to rescue an agent of the archdevil Zariel, who has been captured by the Absolute. You can take advantage of this to negotiate a reward, such as the end of Wyll's pact, or a magic item. Many players were worried about not finding it while progressing in Chapter 2, and this is normal, since it is only possible at the very end, just before its conclusion. You must pass through the Gauntlet of Shar, kill Balthazar, and confront Kéthéric Thorm at the top of the Towers of Hautelune. Once in the second part of Oubliettes, you will end up finding the famous she-devil: Miroza is trapped in an Illithid capsule.

It is absolutely necessary to free her. If you kill her, Wyll will also disappear, in a particularly abominable way. Not releasing Mizora before the second encounter with Kéthéric Thorm has the same result. Remember to search the area carefully before concluding the chapter. If you can't kill Mizora, you can always make fun of her.

Chapter 3: Mizora's Ultimatum

Chapter 3 is also rich in events linked to Wyll. You can speak to Mizora, and even the Grand Duke before and after Gortash's coronation. Unless you kill Gortash during the ceremony, which is strongly discouraged, new events are brewing

A few nights later, Mizora will arrive at the camp and impose an ultimatum on Wyll

- Or regain freedom with the end of his pact upon the death of the Absolute, in exchange for the almost assured death of his father.

- Or become eternally a slave of hell, in exchange for which his father will be saved.

The choice is apparently difficult. Although it's not a very well-received option at first, it's better to choose to free Wyll from his pact, you can save the Grand Duke on your own. Your choice will have important consequences on the end of the game concerning Wyll, but also Karlach. At least, if she's alive.

Save the Grand Duke

If you did not kill Gortash during his coronation, the Grand Duke is transferred to the underwater prison of the Iron Throne. You can find all the details in the guide below. But to summarize the important points, you have to reach the submarine hangar via the sewers or the Fkymm warehouse.

You must use the submarine to reach the prison, and save the Grand Duke before the area explodes. It is advisable to send two characters to open the Duke's cell, in order to heal him, or even teleport him to safety. When it's his turn to play, Mizora appears, forces him to kneel for a turn, and causes explosive spiders to appear. Use the "Help" action on the Grand Duke, treating him, or even teleporting him, allows you to save his life and evacuate him. Once back at camp, thanks to the power of the larvae, the Grand Duke will finally be able to learn the truth about Wyll's infernal pact, which will lead to reconciliation.

Wild, nouveau Grand duc ou l'aventurier lame d'Avernus?

If Wyll's infernal pact is broken, the Grand Duke decides to make Wyll his successor, and offer him his place after the final battle. As always, you can influence their decision. Both options mainly have "role play" consequences which only really come into play during the conclusion, with the different endings of the story. If Wyll remains an adventurer, he becomes the Blade of Avernus. This allows him to continue his adventures with you, if you are a couple, or he will continue them with Karlach, in order to save his life. This unlocks this additional ending for Karlach, which is significantly more positive than the others.

Conclusion and possible ends

This is not an exhaustive list and there are different variations, but here are the different endings awaiting Wyll:

- Dead or transformed into a Lemure in Hell

- Condemned to serve the underworld during his lifetime, and to become a devil after his death

- Grand Duc

- Blade of Avernus, who continues his adventures in the company of the protagonist or Karlach

GUARDIAN ANGEL

Creating a character in Baldur's Gate 3 can already take a lot of time, between choosing the class, race, and appearance, among other things. But players who didn't participate in early access must have been surprised to have to configure the appearance of a second character described as your "Guardian Angel". Although to a lesser extent, it must have also come as a surprise to veterans, since this facet of the game changed with the release of the full game.

Explanations without spoilers

The Guardian Angel is a character who will appear later during the adventure. You don't need to worry about it right now. It's not something you're likely to miss or forget along the way. It is not a companion, nor an invocation. It's also not like the pawns in Dragon's Dogma. It is advisable to give it an appearance that you like, or even find attractive, for immersion reasons. It's best to save the fanciful choices (hideous character, or identical to your protagonist) for later.

What is the Guardian Angel? (Spoilers)

From a certain point in the adventure, the Guardian Angel will appear in your dreams, as well as those of other members of the group. He (or she) reveals to them that he saved their lives when the nautiloid crashed, and he watches over them from what appears to be a supernatural battlefield. It encourages you to use the power of the larva, in order to survive, and to defeat the forces of the Absolute. He will also encourage you to use the larvae extracted from awakened Souls serving the Absolute.

This seems contradictory, however, and it is impossible to determine at first whether he is a benevolent but pragmatic character, or whether he is a trap set by a very powerful dream entity, like the Absolute or others, in order to corrupt you. .

There is absolutely no doubt that the way you treat this character, and whether or not you decide to use the power of the larvae, will heavily influence the end of the game. The guide dedicated to these choices and their ramifications will arrive soon we cannot say more at the moment.

For the record, this character was very different on early access, he wore a light dress and openly made sexual advances towards you, while encouraging you to use the larva. Which made the possible trap much more obvious in this context. Larian Studios may have anticipated all this to put players on the wrong track.

If you really want to know everything about the Guardian Angel's real identity, his potential romance and more, check out our guide to the Emperor.

WHERE TO FIND KARLACH

The first companions of Baldur's Gate 3 are recruited quickly, easily, and in succession, even if they do not necessarily have very good character. But things will quickly get complicated with Wyll and Karlach.

Start the quest

At the start of Chapter 1, after waking up on the beach, you can advance in the area, until you reach a hill, and the entrance to the Druid Grove, the Emerald Enclave. A battle will be fought between the goblins and the defenders, including Wyll (if you do not play him directly). It is important to keep Wyll alive, even if his chances of dying are low.

After entering the Enclave, you can talk to Wyll, who is training children. He asks you for help in finding a Devil, in exchange for which he agrees to join your group (or your camp).

Where to find the Devil?

The search area is displayed on the map as a large yellow circle. You shouldn't have any difficulty finding your way around.

The normal path is to go through the Dilapidated Village, in the center of the map (watch out for the goblin ambush), then take the large stone bridge to the north. Then follow the road to the east (watch out for the ambush of hyenas and gnolls). Continuing to follow the road going down towards the river, you will end up coming across corpses, then a trunk serving as a bridge to an islet, on which Karlach is located. She's the "devil".

If your goal is to complete this quest or recruit Karlach as quickly as possible, you can take a shortcut through the woods, going north before entering the village. You can cross the river by jumping on the rocks. This makes it possible to reach Karlach without the slightest fight.

Dialogue with Karlach and possible choices

The discussion with Karlarch can take several different turns, but the most determining element is whether or not Wyll is in the group. Remember to save before approaching.

- If Wyll is absent, you can easily calm the situation and recruit Karlach, if you agree to go kill the false Paladins of Tyr, present in the garrison above the cliffs.

- On the contrary, if Wyll is present, it is a little more difficult to disarm the situation, but it remains possible with skill rolls or good response choices.

It is also possible to simply kill Karlach, in order to satisfy Wyll. But both choices will have repercussions on the latter's personal quest.

The Paladins of Tire

Upon reaching the garrison full of gnoll corpses, a little above the area where Karlach is located, you can come across a group posing as Paladins in the service of the god of justice. If Karlach is not in your party, they will offer you a holy sword as a reward, if you agree to kill her. Clearly, a lot of people want her head, poor thing.

If you go to talk to them with Karlach, they will ask you to kill her. During the dialogue, you can try to determine if they are really Paladins of Tyre. It is possible to detect that the "Paladin" is lying, with a perception roll. Before speaking to their chef, you can also use the Communicate with the Dead spell on their deceased companion in the kitchen. He will reveal you to be an impostor, and that they all serve the she-devil Zariel. This means that the "right" choice is to kill these impostors, and defend Karlach.

In any case, following dialogue, the fight will begin, either with Karlach or with the rather formidable group of false Paladins. It includes a Magician, a Thief capable of turning invisible, and a fearsome Paladin with a great sacred sword.

If you kill Karlach, you lose a companion, but you receive the Great Sword of Tyr (+1). If you kill the impostors, you recruit Karlach, and you get all their equipment, including the great sword. This second choice is obviously much more profitable.

Wild vs. Karlach

If you recruited Karlach in Wyll's absence, a confrontation will take place at the camp during the next Long Rest.

Its flow is quite similar to the dialogue that takes place if you meet her with Wyll. You can try to convince Wyll to abandon his mission with skill rolls and good dialogue choices. It is also possible to kill Wyll or Karlach, depending on the circumstances.

By conducting the negotiations well, Wyll will accept the presence of Karlach, and recognize that he has been manipulated. This will nevertheless have repercussions, since a few days later, Mizora, his demonic mistress will appear in the camp to punish him, which will alter his appearance.

WITHER

The adventure in Baldur's Gate 3 wouldn't be the same without a friendly undead with divine powers squatting your camp. To begin with, it would be significantly less comfortable.

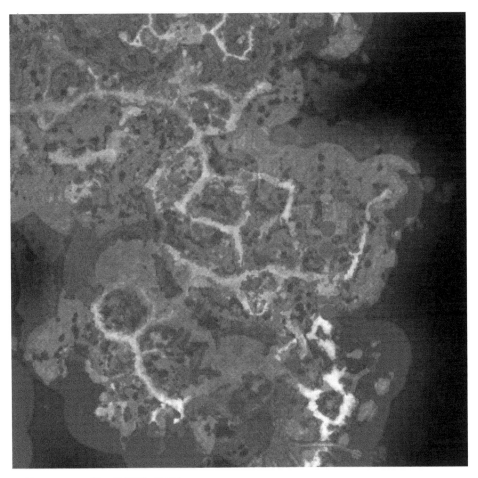

Where to find Blight?

When you know exactly where to go, Wither can be recruited within minutes of completing the tutorial. Indeed, the door at the end of the beach serving as the starting point of chapter 1 leads directly to the Damp Crypt in which he resides. You can pick it or destroy it, but it's better to recruit a few companions first. You can also take the path above the cliff, which leads to several entrances, after killing or scaring away the grave robbers. You can convince the doorman to let you in, or as a last resort, make the rock fall from a height by pulling on its rope. This will open a passage.

Once in the Damp Crypt, reach the large room at the back of the area. Next, look for the hidden button to open the secret room. After confronting the undead guards, open the sarcophagus to free Blight. Your response choices during the dialogue have no importance. Just know that he will join your side. Even if you attack him, or if you refuse to answer.

If you ever missed the crypt, there's no need to worry. He will simply invite himself into your camp a little later.

Usefulness of Wither

This brave undead refuses to answer your questions, and even to die. He will praise you even if you attack him, since he won't take damage. On the other hand, its services are really useful. You can always find it in a corner of the camp, remember to open the doors when you are in a building.

It offers the following services:

- **Respec** (100 Gold) - Allows you to return a character to level 1, which allows you to modify almost everything apart from their race and appearance. The starting class and characteristics can be changed. This is very useful for your original companions, who tend to have atrocious traits, or the wrong class to accompany you. The experience is retained, you can then win the remaining levels.

- **Resurrection**(100 Gold) - Whether due to combat, or an unfortunate fall into a bottomless precipice, death is never far away in BG3. Even if only a character's soul remains, and even if it's your protagonist, their companions can bring them back to life in exchange for a modest fee.

- **Servants**(200 Gold) - If you don't like the basic companions, or they have disappeared for one reason or another (monster!) you can replace them with freely configurable characters. This ensures you have exactly the band of your dreams. On the other hand, these characters don't really have any personality, and don't expect to have a romance with them.

A fun detail is the ability to pick Blight's pockets. All the gold you spent on his services remains in his inventory, and you can steal from him. Even if he spots you, he doesn't say anything. In summary, it's completely free.

What is Blight's real identity?

Spoiler: It's never said clearly, but the clues are so numerous and not very subtle that there is almost no doubt about the identity of Wither. Everything indicates that his real name is Jergal, the ancient god of death, who offloaded his responsibilities to the three dead gods (Bhaal, Baine and Myrkyl). He now works as a simple scribe of the dead under the current god of death, Kelemvor. One of the books in play is dedicated to Jergal, and it is written that he asks the same question to everyone he meets the first time he meets them: "What determines the value of a person's life? mere mortal?" Blight asked you the exact same question. There is no wrong answer.

Playing the original Dark Urges, he has a special role to play in the Temple of Bhaal Otherwise, it only intervenes very little in the story. He guides Arabella for a while, at least if she survives and you find her parents. We learn more about his motivations during a special cutscene, after the end credits of the game. He then mocks the Three Dead Gods, whose plan was to capture a large number of souls by transforming the residents of the Sword Coast into Mind Flayers. But their plan obviously had quite a few flaws, since the souls are in reality completely destroyed during the transformation. As a scribe of the dead, Jergal/Blight could not authorize something so nefarious, hence his indirect intervention: he assists you during your adventure, with powers far beyond what mortals have access to .

SAVING ALFIRA

Alfira is one of the notable characters of the first chapter of Baldur's Gate 3. This bard lacking inspiration has long been suspected of being one of the original characters to come for the release of the game. Ultimately, she is not the case, quite the contrary. Ironically, or out of cruelty, the developers even made him an almost obligatory victim of the last original character of the game: Dark impulses. You could say he's killing her metaphorically and literally.

Where to find Alfira?

This bard is found in the Druid Grove, on top of the cliff to the east of the central part. She plays music in front of a group of animals. (X281:Y:494). For the record, using a spell or a communication potion with animals to speak to your audience is quite amusing.

Help Alfira finish her song

It's not a real quest in itself, but you can suggest different lyrics for his song, and encourage him, in order to help him grieve. The Bard class is particularly adept at this, but it is far from mandatory.

- Don't leave the conversation, don't tell him to give up.
- Offer to help when you can.
- Respond positively when she asks you about the end result.

If you choose to eliminate the goblins, afterward, she will head to your camp to party with the rest of the Tieflings. She will appear again, later in the story.

Dark impulses and Alfira

If you've decided to pick up the original Dark Impulses character, you should start to realize that this title isn't just there for show. Despite your best efforts, blood flows steadily. There is a special event related to Alfira in this area, since, if you have inspired her at the grove, she offers to join your camp and spend the night there. This event does not occur with other origins.

The problem obviously being that if you accept that Alfira joins you, Dark Impulses will kill her in her sleep, during your next Long Rest. The next morning, a character comes to give you a reward for your action: a magnificent cape offering mastery of Discretion. It also makes you invisible for 2 turns, each time you kill an enemy. It's a nice reward, but if you don't want to sacrifice Alfira, you'll have to get creative.

Method to try to save Alfira

Players have discovered unorthodox methods to spare Alfira. Here are the steps to follow scrupulously:

● To start, refuse to let her join your side. Send her back!

● Kill your Dark Urges protagonist at the camp, using attacks from your companions for example.

● Control one of your companions, then spend the night.

● Bring your Dark Urges protagonist back to life with a scroll of resurrection.

Go talk to your Butler Sceleritas Fel. He'll make some comments about how Alfira was handed to you on a silver platter for you to kill, and it proved much harder for you to spare her.

However, it seems that with this method, your butler still assassinates Alfira for you. The advantage of this method is that at least you don't break your oath as a Paladin.

Alternative method to really save Alfira

As the method listed above did not prove sufficient, players found an even more twisted method to save Alfira.

● When you initiate a Long Rest and the cutscene during which Alfira joins the camp is triggered: Load your last save. Preferably just before activating Long Rest.

● Go see Alfira then without activating Long Rest. Then knock her out using a blunt weapon and non-lethal damage. You can select the option in the left interface. The pommel of a weapon also does the trick, as long as it drops it to 0-1 HP. You can try to do this by approaching discreetly behind his back, in order to avoid going into combat and losing your relationship.

● Confirm that she is indeed unconscious, and not dead.

● Another character will show up at the camp to spend the night: Quill Will. And he will become your victim instead of Alfira, which allows you to receive the cape as a bonus.

● If you don't want to kill him either, use the method listed above: Send him away, kill your protagonist, spend the night, then bring your character back to life. Quill Will is killed by the butler, but Alfira will survive.

The big flaw with this method is that there is a good chance that Alfira will hate you after attacking him.

SAVE VOLO

Even if he is less known than Elminster and Drizzt, Volo is one of the most emblematic figures of the Forgotten Realms, since he is the author of numerous works. It's a bit like merging Dandelion from The Witcher and Deckard Cain from Diablo: a jester who always needs to be saved, but who also has great knowledge. He made an anecdotal appearance in the first Baldur's Gate, and he is given a more important role in Baldur's Gate 3.

First meeting with Volo

In the majority of cases, you will meet Volo for the first time in the Druid Grove. He is in the central area, talking to a bear. You can talk to him to find out more about his presence here, but there is nothing special to do at the moment.

Save Volo

You will find (or meet) Volo at the entrance to the Goblin Camp. He has been captured by these gremlins, and he is forced to distract them by singing. You can choose various answers, or make skill rolls, but he will be taken by force into the inner sanctum of the base.

- As you enter the sanctuary, explore the rooms on the right. You will eventually come across Volo, locked in a cage. He asks you to help him escape.

- Talk to the goblin next door, you can trick or intimidate her so that she accepts that you simply free Volo. You can also kill her and open the cage, if you're in a bloody mood.

- Open the door to Volo's cage, and meet him at your camp. You can find it there

Remove the parasite

By returning to your camp, you can speak to Volo, who will give you a Bard outfit to thank you (Light armor, 11 AC, gives life on bardic inspirations). He will also temporarily serve as a merchant in your camp.

You can talk with him, and reveal your situation with the larva in your head. He suggests you remove it with a needle and his expertise.

What follows is a scene that is at once hilarious, grotesque and painful, during which Volo will try to remove the larva. You'll regularly have the opportunity to interrupt the sequence, as it quickly becomes apparent that Volo doesn't know what he's doing. But if you decide to go through with it, he ends up gouging out your eye. This also causes a slight loss of approval among the majority of your companions.

As an apology, Volo will give you a magical prosthetic eye. This will give you minnow eyes, as well as detection of invisible creatures within 12 meters. There are no other penalties, so it's a good choice, both to get a good laugh (or cringe) and receive a nice bonus. Volo will then leave the camp and disappear for the moment.

241

During your journey in Baldur's Gate 3, it is relatively common to encounter allies capable of helping you on one or more occasions in order to facilitate a fight that would normally be too difficult for you. One of these encounters takes place in Act 1 and puts you in contact with three unusual ogres, to say the least.

Meeting the ogres of Baldur's Gate 3

During Act 1 of Baldur's Gate 3 you will meet many more or less interesting characters, some more friendly than others. One of these encounters that is surprising to say the least, even frightening, is the one that you have sooner or later with a trio of ogres in the Devastated Village, near the Goblin Camp. But while the Heroic Fantasy universes have often taught us that ogres are somewhat stupid and above all hostile beings, here it is not the case... Well, not quite.

As soon as you approach these ogres, a discussion will automatically start. It's Lump the Enlightened who is conversing with you, he is probably the only truly intelligent one in the group thanks to a powerful artifact that he has in his possession (one of the best in the game in fact).

Initially, Lump describes you to his friends as "their future meal", unless you have the Mark of the Absolute. This can be obtained from Haruspia, a goblin priestess devoted to the Absolute found inside the Devastated Sanctuary, in the heart of the Goblin Camp. If you have accepted the Mark, you can present it to Lump and he will no longer threaten you because he will consider you an ally. But if you don't wear it, then several possibilities are available to you:

- **[Without risk]** "What eloquence for an ogre!" : A few exchanges take place with Lump, after which the other choices are offered to you again.

- **[Failure]** "I bear no such mark": Lump then considers you an enemy and invites his friends to feast on your carcass. A fight with the three ogres begins.

- **[Risky, Recommended by default]** "I have even better: I am one of the chosen ones of the Absolute": If you succeed on a dice roll of 10 or more in Charisma (Deception), Lump and his allies are now convinced that you are an ally for them and offer you the choices mentioned below. If you fail, Lump and his friends attack you.

- **[Risqué, Recommended if you have points in Intimidation]** Dialogue varies depending on your class: If you succeed on a dice roll of 10 or more in Charisma (Intimidation), Lump and his friends resign themselves to considering you as an ally after your threat and offer you the choices mentioned below. If you fail, all three attack you.

- **[Failure]** "Attack": A fight begins immediately with the three ogres.

Your options are quite limited, but if you succeed then the rest of the dialogue will perhaps lead you to obtain these three ogres as formidable allies for the battles that await you. And that's not a luxury!

Recruit Lump, Fank and Chock in Baldur's Gate 3

If you managed to get through the first part of this interaction with Lump the Enlightened, five new choices are available to you. And here again, you will have to be careful with the words you use or the actions you undertake otherwise you will rob these three colossi:

- [Without risk] "You speak very well for an ogre": A few exchanges take place with Lump, after which the other choices are offered to you again.

- [Without risk] "I thought I saw that you didn't wear the mark": A few exchanges take place with Lump, after which you have the possibility of offering him to put himself at your service, or to exchange a little more with him via the other lines of dialogue.

- [Without risk] "I have encountered very few other ogres in the area": A few exchanges take place with Lump, after which the other choices are offered to you again.

- [Failure] "Attack": A fight begins immediately with the three ogres.

- [Without risk] "Leave": You leave the discussion without the ogres chasing you. You can come back to it later without incident.

After asking the ogre why he wasn't wearing the Mark of the Absolute, you get the chance to ask him to join your band of merry men. But once again, you will have to choose intelligently so as not to ruin yourself for nothing:

1. [Risk-free, Not recommended except as a last resort] "I will give you 500 gold coins to call on you in case of combat": You immediately spend 500 gold coins and purchase the services of Lump and his friends.

2. [Safe, Not recommended unless you call the ogres on a tough fight and let them die] "I will give you 1,000 gold coins after calling on you in case of combat": If you succeed on a dice roll of 10 or more in Charisma (Deception), the ogres agree to join you during your next battles but you will have to pay them 1000 gold after the battle. If you fail, only options 1 and 4 are available next. Important: If Lump and the other two die during the battle you called them for, the payment is canceled.

3. [Risk-free, Recommended as first choice] "I will pay you with the flesh of the vanquished... You will have something to feast on": If you succeed on a dice roll of 20 or more in Charisma (Persuasion), the ogres agree to join you during your next battles without spending a single penny. If you fail, only options 1 and 4 are available next.

4. [Without risk] "I changed my mind, forget it": You leave the discussion without the ogres chasing you. You can come back to it later without incident.

The ideal is to follow choice 3 in order to buy their services without any cost for sure, and in case of failure you fall back to choice number 1 which remains a relatively acceptable alternative. If, however, you know that you will face a really tough enemy and that the ogres will die during the battle (or that you will help them die a little...), option number 2 is undoubtedly the most desirable to reduce the costs !

Once the services of these three mountains of muscles have been obtained, you then obtain the Lump Horn in your inventory. When you use it, the three ogres arrive on the battlefield and wreak havoc by massacring your enemies. The trio is truly extremely powerful (during Act 1 at least), and you can call on them as many times as you want as long as at least one of the three comes out of the fight you're using them for alive. you called!

The other excellent news which justifies the price of the services of these three greedy ogres is that when they die you can recover the loot in their possession from their corpse. And although Fank and Chock have nothing really interesting, you will be able to collect the Warped Headband of Intelligence which will increase the statistic of the character who wears it to 17, a really very, very powerful helmet if you use it correctly. And believe us: for 500 gold coins you won't get anything better for a while!

SAVING HALSIN

It's generally easy to recruit the original characters into your party in Baldur's Gate 3, but they aren't your only options. There are a number of side companions, which require more effort. Halsin the Druid is the most notable example, especially after the publicity that Larian Studios gave him on stream, in his bear form.

Save the first druid

The first step is to go to the Druid Grove, then talk to Rath, after the confrontation with Kagha. He asks you to save Halsin, the leader of the druids, who has gone to the goblin base, to the west of the map. There are several ways to infiltrate the premises, for example by saving Sazza the goblin. That's a lot of rescues. In any case, you have to cross the Ruined Village, then the bridge, the Goblin Outpost, and finally, the party in front of the Goblin base. You can use infiltration, or different skill rolls to convince them to let you pass.

Once in the base, go to the back right of the area to reach a door leading to the Worg Enclosure, the equivalent of the kennel, and the goblin prison. In the cage on the right, you will come across a group of children harassing a bear in a cage. It's Halsin himself, and he's going to ask you to help him escape. Of course, it's better to agree immediately, and a battle ensues. Kill all the children before they have time to escape (this game doesn't joke), otherwise they will sound the alarm and reinforcements will arrive.

After the battle, talk to Halsin, and he asks you to kill the 3 goblin leaders: the Goblin Priestess of the Absolute, the Hobgoblin Warchief, and the Drow Paladin

Minthara. You can choose whether or not to be accompanied by Halsin to ravage the base. If you refuse, he will simply return to the Druid Grove, waiting for your return.

Kill the goblin leaders

What happens next involves shedding a lot of blood. You can go about it in more or less subtle ways, but you have no choice if you want to recruit Halsin. You must kill the 3 designated targets, which means that you will not be able to recruit Minthara the drow. It's mutually exclusive with Halsin's recruitment, as far as we know.

Minthara is in another room all the way to the left of the base. The High Priestess is in the entrance hall, and it's easy to get her to speak to you privately in a secluded room. The biggest fight is against the warlord, in the throne room. Attacking him from the high beams is a good way to gain the advantage. Once the 3 leaders are killed, return to the Druid Grove. If Kagha is alive, you will come across a dialogue between her and Halsin.

By speaking to Halsin, he will direct you towards Rath to receive your reward: the wolf rune.

Return to talk to Halsin, he will let you know that he intends to travel with you to lift the curse of shadows at Highmoon Towers. Go talk to Zevlor at the grove, then again in front of the entrance, to organize the Tieflings' party in your camp, during the night. Halsin will also be there, which will be an opportunity to socialize with him again. Unfortunately, Halsin is not joining your group at the moment, you will have to wait until you have progressed in chapter 2 for that.

Chapter 2: The Final Glow

You can advance to the next chapter of the story by entering certain areas: The Mountain Pass from the first region on the surface, or the Highlune Woods, from Maleforge in Outland. Bring Light spells and/or torches. Your new objective is to reach the inn, the Last Glow, more or less in the center of the region. It is easily recognizable by the immense dome of light that surrounds it. An old acquaintance of Baldur's Gate players awaits you there, the druidess Jaheira, even if that is not what interests us in this case.

Waking Art Cullagh

Enter the inn, then enter the first room on the right. An injured man named Art Cullagh mutters incoherent words on the bed. You are asked to examine him to determine his illness. Many skill rolls are possible, as well as possible interactions with the different classes. You can also read his thoughts with a spell.

Return to your camp, then talk to Halsin about this man. He has a connection to the shadow curse, and he whispers an evocative name: Thaniel.

Halsin is now at the injured man's bedside in the inn. He will share his conclusions with you: you must go to the Healing House, not far from the Towers of Hautelune.

245

The position of this house is indicated on the map.

The healing house

Before attacking the rest, you will need the blessing of the priestess of Selune on the upper floor of the inn, which will trigger an attack. Then, you will have to complete the quest to obtain a way to protect yourself from the darkest shadows, by attacking a patrol in the company of the minstrels. You'll need the Moon Lantern in his possession (or the pixie inside) to venture into town.

Head to the Healing House next. In the central part are the surgeon and his 4 assistants, Doctor Malus Thorm. You can try to convince him to sacrifice his assistants and then kill himself, if you have the right class and excellent skill rolls. Otherwise, you will have to confront him and kill him. On his corpse, you will find a damaged Lute, which you will have to pick up.

Saving Thaniel

Return to the Last Glow Inn, then speak to Halsin. He will ask you to play a few notes on the damaged Lute, which will wake the man in a coma. After a dialogue, Halsin will meet you at the edge of the lake, north of the inn.

He will ask you for your help to save his friend Thaniel, by guarding a magic portal. He will venture inside alone, and you must protect him for 3 turns against waves of enemies. It doesn't seem like much, but literally dozens of enemies will appear, a real army. It is better to have taken a long rest beforehand, and prepared spells, invocations and consumables. Your spellcasters will also have fun: the fireballs and hailstorm will wreak carnage.

Recruit Halsin

If you manage to protect the portal all the way, Halsin will emerge with Thaniel. He will then meet you at your camp. Once there, you will notice that Halsin finally has his own tent. He finally offers to join your group, and to assist you until the end of your quest if you are interested. You will finally be able to improve your relationship with him, ask personal questions, and help him pass his levels.

To complete Halsin's quest line and ensure that he remains in the group, you must lift the curse hanging over the region. By talking to Halsin and Thaniel, you will get clues. You must venture into the cursed lands to find Oliver, a corrupt child asking you to play hide and seek with him.

After winning the game, you can ask him to merge with Thaniel. If you refuse, or if you fail during the hide-and-seek game, you will have to confront him and then convince him to merge with Thaniel.

By completing all of these steps, the curse will be lifted when you leave for Baldur's Gate at the end of chapter 2.

In Baldur's Gate 3, devils love to make deals with mortals in order to obtain their souls, and they understand that being attractive makes things a lot easier. Raphael is a good example, as is Mizora. The latter can only catch the eye, and it is possible to have some unexpected interactions with it. But be careful, she could well turn out to be your worst enemy: expect her to leave with your companions, the camp, Gratouille and your soul in the event of divorce.

But first, here's how to get the Sex Scene with Mizora, which involves making the right choices during the adventure. Spoiler alert.

Effects of romance with Mizora, on your other relationships

Talking about romance is obviously questionable with a devil who is after your soul But either way, the chance to spend the night with Mizora doesn't come until pretty late in the story, and you probably already have one or more romances in the works at that point. As in many RPGs of the genre (and real life), the majority of companions don't want to share you with other people. While maintaining a romance with Karlach and Shadowheart becomes impossible after a certain point, forcing you to choose, this is not the case with Mizora. Your romantic interest seems to consider it just a one-night stand, a bit like going to the brothel in Baldur's Gate, although there is a little dialogue tied to it. We could not confirm whether all characters have such a detached approach. Wyll doesn't like it, though, and he's going to disapprove, which is pretty logical. If you have a romance with him, he may take it quite badly.

How to start romance with Mizora?

❖ Chapter 1

The first step involves rescuing the Tieflings from Druid Grove. Wyll leaves your party if you team up with Minthara and the goblins. You can choose to help and recruit Karlach or not. When Mizora arrives at camp a few days later, your choices don't seem to matter too much, although it can't hurt to be nice to her and start charming her.

❖ Chapter 2

You must then reach the second chapter of the game, by entering the Mountain Pass or the Shadow Woods around the Towers of Hautelune (via Malforge). During a night spent at camp, Mizora will show up to ask you to help a she-devil from Zariel who is trapped in the Absolute. You must accept. You can negotiate additional terms if you wish.

You will not have the opportunity to do this before the very end of chapter 2. You must have killed or freed Chantenuit, then confronted Ketheric Thorm on the roof. He will then flee into an Illithid nest under the tower. While exploring the northern part of the area, you will meet Mizora, prisoner in a Mind Flayer tank. Turns out she was the she-devil who needed help.

Inspect the consoles next to it, you must either choose the option to open the tank, or use [Force] directly, if you are afraid of making a wrong move, or if you prefer to avoid using your larva.

❖ Chapter 3

The rest doesn't happen until chapter 3. Cross the bridge to arrive at Wyrm Rock, where Gortash's coronation will take place. He invites you to participate via one of his Steel Guardians. As you advance in the area, you will also come across Mizora. It is advisable to have Wyll in the group, in order to be entitled to some additional dialogue, and potential approval bonuses. During the coronation, avoid the fight, it is lost in advance. Talk to Mizora again after the ceremony.

The following night, Mizora will arrive at the camp again, in order to impose a painful choice on Wyll: sell his soul in perpetuity, or dissolve the contract, but his father is almost certain to die. You can influence Wyll however you like, it has no impact on the potential romance with Mizora.

The important thing is that Mizora decides to stay at the camp afterwards, which allows us to talk with her. You shouldn't chase it away, of course.

Get the Sex scene with Mizora

The rest comes after a few nights. An exclamation point will appear above Mizora's head. She will then offer you the opportunity to taste infernal pleasures, well beyond what the primary plane can offer.

It goes without saying that you have to accept, since that's what you're there for. But we advise you to save the game before, in order to resume it later to refuse, if you are afraid that there will be negative repercussions on your main romance and your relationships. You can also stop everything at any stage by choosing the last option, if you find the scene uncomfortable.

The following choices are pretty obvious:

- "I like it a lot, what do you suggest?"

- "Immerse yourself in the feeling.", "I'm ready.", "Prepare yourself for the worst." "Drink you from the river of blood."

- Then, Mizora invites you to discover some of the hells, with sensations linked to them. This also influences the visual effects of the erotic scenes that follow.

- Finally, once Mizora has undressed, you can choose to approach her and nestle between her wings, which will trigger the final scene.

Consequences

You will speak to Mizora the next morning, and your companions will have comments to make on the subject. Especially the character you have a romance with. Shadowheart doesn't seem jealous, but we can't guarantee that this will be the case for all the other characters, and that there won't be long-term consequences, for example, on the end of the game. If in doubt, you can load your game.

And if you doubt it, everything indicates that Mizora especially has the night with you to corrupt your soul, and sow trouble in your camp.

MINTHARA

It's generally easy to recruit the original characters into your party in Baldur's Gate 3, but they aren't your only options. There are a number of side companions that require more effort. Minthara the Paladine is the most notable example, a valuable ally that you were far from suspecting during your first possible meeting with her.

September 7 Update: With the latest update of the game, Minthara is entitled to much more dialogue and interactions with the protagonist and other characters. This makes for a more interesting romance and companion in the past.

Chapter 1: Meeting with Minthara

Your first meeting with Minthara takes place during Act 1. While you are missioned by the Tieflings and Druids of the Druid Grove to ward off the coming invasion of the goblins from the West, you finally discover that behind these little beings treacherous greens hide three great leaders who turn out to be awakened Souls: Haruspia, Dror Ragzlin and Minthara who is the one who interests us here.

● **Important** : Killing Haruspia, Dror Ragzlin or any goblin in Act 1 can jeopardize your plans with Minthara. If you have to kill them, it must be done through dialogue options, you must not attract attention and alienate the entire goblin camp!

By going to meet her inside the Devastated Sanctuary, a secluded place in the heart of the Goblin Camp, you discover that she plans to storm the Druid Grove with the help of an army. search for a powerful artifact that she thinks she will find there.

To ensure you make Minthara an ally, you must at all costs share her ambitions to purge the Grove of any druidic or tiefling presence. This therefore implies opting for what we would call a "Dark Run" or an "Evil Run", in other words doing evil. The consequences of this choice are immense for the rest of your adventure, so be aware of what this could imply.

Assault on the Druid Grove

For Minthara to trust you, you must agree with her when she mentions her imminent assault on the Grove. Do not contradict her faith in the Absolute, do not attack her under any pretext and do everything in your power to lead her to the Tieflings and Druids she is looking for, only in this way can you she will stand by your side.

When she asks you to interrogate the prisoner held captive in the Ravaged

Sanctuary, again be sure to be tactical in your approach to the executioners: they must not become hostile, so just explain to them that they do not have the skills for such interrogation and prove yourself by torturing the poor man until he reveals information about the Druid Grove in the presence of the two torturing goblins. Once this task is accomplished, your two new "friends" will report to Minthara who will then congratulate you on your "sense of persuasion".

● **Important** : You can also save Volo nearby in the process, but again you will have to act intelligently since his captor must not die or become hostile towards you.

As soon as you have reported to Minthara about the rigorous interrogation you have just carried out, she will immediately begin preparations for the assault on the Grove. At the same time, she will instruct you to go inside the village she plans to assault and sound the horn near Zevlor (above the main gate) in order to start the battle. Take the opportunity to take a long rest, the Paladine told you that the attack would be launched at dawn.

Once inside the Grove and after a good night's sleep, Zevlor asks you for information regarding your infiltration of the Goblin Camp. Keep playing his game and pretend you're going to save everyone and decimate the goblins to protect them. He will then be reassured, after which you can click on the nearby War Horn to start the final battle.

It is at this moment that the bulk of your journey takes place. When the goblins arrive with Minthara at their head, she asks you through the Illithid larva to kill Zevlor and open the door to allow her troops to enter the city and wreak carnage. You only have one chance: "Obey and open the door." Zevlor then panics, and you must fight against your ex-allies with an army of goblins, bugbears and other spiders and ogres to support you.

● **Attention** : Turning against Zevlor and his people will attract the wrath of Wyll who will attack you instantly if he is in your group. This will make him disappear permanently from your little company.

After having defeated the defenders of the gate, all you have to do is go to the villagers' hideout in the Grove to massacre the last survivors who will not represent the slightest danger to your new army. Minthara congratulates you at the end of this carnage and tells you that she will join you in your camp at nightfall.

Romance with Minthara

Back at your camp, the goblin army beat you to it and started the festivities. Minthara is also present, obviously, and makes you an unexpected proposition to say the least, believing that this night "you will belong to her". Simply interact with the Drow in your camp to initiate the rest of the events by confirming your wish to spend the night with her.

Then go to bed to have access to several possibilities depending on your current romances. Ignore those at your disposal and select "Minthara" to reach the

Paladine a little further away, out of sight. During the rest of the events, just comply with Minthara's requests so that the romance evolves throughout the night, obviously making sure to never reject her advances.

After spending part of the night in her company, you discover a more vulnerable Minthara snuggled up against you near a campfire. You then have several possibilities, and we strongly recommend that you choose the option inviting you to "concentrate on your desire, only on your desire" to avoid any disastrous dice roll (a 15 or more is necessary for the dice roll). Wisdom).

The elf then wakes up and invites you to open your heart to him, opt for the option you prefer, it has little importance for the rest of the events except that it will reveal things to you about her in return. On the other hand, when the artifact subsequently begins to react to protect you from Minthara's psychic intrusion, be sure to remain faithful to the Absolute at all costs and above all not reveal that you have the artifact that she is looking for.

After that, during your night your new ally will be seized by a murderous impulse and will try to assassinate you in order to find the voice of the Absolute which rocked her until your meeting with her. To get out of this situation, make a Persuasion roll to meet this supposed goddess in the company of Minthara (a 2 or more is necessary).

Chapter 2: The Judgment of Minthara

After that, you finally see yourself as allies and she sends you to the Hautelune Tower at the heart of Act 2 of the game, inviting you to go through the Mountain Pass, a choice that we recommend and which does not prevent you from discovering the Dark Depths as long as you do not pass the Temple of Shar stage during Act 2 (an in-game warning warns you in the event of a point of no return to Act 1).

You won't hear from Minthara again until the near-denouement of Act 2, so you can proceed as you wish until you reach Highmoon Tower, the culmination of this second part of your adventure. . When you first enter the tower, you find that your Drow ally is being judged by Ketheric Thorn himself. Your only option at this moment is to silently observe the scene in order to avoid any problems for both you and her.

After the judgment, Disciple Z'rell sends Minthara to the heart of Highmoon Tower Prison, to the basement of the tower. After a short cutscene, you have the choice to leave the surrounding goblins alive, or to execute them, it is up to you to choose this has no impact on the rest of the events. Then go to the prison via a hidden staircase in the eastern part of the tower to find your imprisoned ally.

In the prison, you discover that Minthara is undergoing an interrogation far more violent than the one she encouraged you to subject to the prisoner in Act 1. In fact, she is attacked psychically by two torturers, Jasin and Sumera , causing her terrible brain pain with the aim of formatting her psychically. Close the door behind the door so as not to alert the guards in the course of events, and then choose what you decide to do, bearing in mind that to recruit her you must at all costs intervene in the abuse she is suffering. , one way or another. Four possibilities are available to you:

1. **[Risqué]** Make it appear that you have been tasked with taking care of this prisoner. Requires a minimum Persuasion die roll of 12 or Intimidation die roll of 15 to succeed. If you succeed, you enter Minthara's mind and another die roll requiring a Persuasion dice roll of at least 18. If you fail, you must choose between killing the torturers or watching Minthara be psychically destroyed by Jasin and Sumera (you must choose the first option).

If you succeed again, you inflict the psychic harm on Minthara yourself. If you fail, you must choose between killing the torturers or watching Minthara be psychically destroyed by these two (you must choose the first option).

2. **[Failure]** You rejoice and contemplate the scene of torture. Minthara is then subjected to Jasin and Sumera's torture without you intervening, ultimately leading to her brainwashing.

3. **[Without risk]** You ask the torturers if doing this is useful since Minthara could be useful. This triggers additional dialogue, ultimately leading to a choice: let her be psychically destroyed, or kill the two torturers to free her (you must choose the second option).

4. **[Recommended, Risk-free]** Call the torturers cowards. This automatically triggers a fight with them leading to their death and the release of Minthara.

After rescuing Minthara from her prison, you have to get out of the cell, obviously. In doing so, a dialogue is triggered with one of the guards who questions you about your legitimacy in releasing the prisoner. You can try to persuade him, or attack him directly which triggers a fight with the guards nearby as well, be sure to be prepared accordingly. Warning: If you attack the guard, you must be hostile towards all creatures in the Highmoon Tower who will attack you on sight.

The escape from the Hautelune Tower

● **Important** : Do not use Fast Travel to leave Highlune Tower during this stage Minthara is not your companion yet, you must first get her to leave the tower. If you use Fast Travel, you will lose access to Minthara permanently.

After your escape, the objective is to get Minthara out of this huge quagmire. When leaving the prison, you automatically find yourself addressed by Adept Merim who tells you that General Ketheric has requested that the prisoner's spirit be reconditioned psychically, not released. Three possibilities are available to you, select the one with which your main character has an advantage:

1. **Persuasion** (Roll of 12 or more): Assure Merim that Minthara is loyal to you and that you are keeping an eye on her. If you succeed, you will leave the tower without problem. If you fail, you enter combat.

2. **Intimidation** (Roll of 15 or more): Threaten Merim by telling him to move away from you. If you succeed, you will leave the tower without problem. If you fail, you enter combat.

3. **Deception**(Roll of 15 or more): Assure Merim that Minthara has been reconditioned as requested by Ketheric and that her will is no longer hers. If you succeed, you will leave the tower without problem. If you fail, you enter combat.

After taking many risks, you can finally leave the Hautelune Tower. As soon as you pass in front of the access portal in front of the tower a dialogue is triggered with Minthara, and you can finally tell her where your camp is located so that she can find you there. And this time, it's not to make you his object of desire but to accompany you in your epic!

JAHEIRA

One of the very first potential companions in Baldur's Gate 1 was the half-elf druid/warrior Jaheira, along with her husband Khalid. She was also back in Baldur's Gate 2, in which she became a potential romance with the death of her husband (that's proper). A century has passed since then, and in Baldur's Gate 3, the years have not been kind to her, which explains why she no longer has the strength of a legendary heroine who faced gods.

Chapter 2: First meeting

As Jaheira is not an original character, you will have to wait until the second chapter of the game before meeting her. It is located at the entrance to the Inn of the Last Light, in the Cursed Lands. It's hard to miss, since it's a huge bubble of light.

For the dialogue to end well, it is better to have saved the Tieflings in the first chapter 1. Be cooperative afterwards, so that she likes you more.

Be careful, when she sends you to speak to the priestess of Selune in order to obtain a blessing of protection against shadows: rest before, and summon the creatures at your disposal. Also make a backup. A big fight will break out, if the priestess dies, the situation will take a disastrous turn. It will be impossible to recruit Halsin, and Jaheira's recruitment looks bad, even if you can still make up for it.

Defeat Kétheric Thorm with the Harpers and Chantenuit

You must then advance in the exploration of the region, and the progression in the story. Go to Shar's Mausoleum and Gauntlet, preferably with Shadowheart if you have confidence in your persuasion rolls. When you reach the final area in Shadowgrey, don't let Shadowheart kill Nightsong, and don't let Balthazar kidnap her either. This would cause a real disaster, leading to Jaheira's outright death. We must free Chantenuit.

Then go to the Towers of Hautelune. If you have managed each stage correctly, Jaheira will be there with his army. If you failed to defend the inn, Jaheira will be alone. In this second case, she absolutely must accompany you, in order to avoid her death.

Progress in the tower, killing the enemies you come across, until you reach the top, and Kéthéric Thorm, who is finally vulnerable. You just have to defeat him, then he will flee. You don't have to do the sequel with Jaheira.

Recruitment

After defeating Ketheric Thorm and the Avatar of Myrkul in the Deep, the surviving NPCs will be found in the Hall of Highmoon Towers. This includes Jaheira, who you can talk to. It is important to invite him to follow you, and to team up with you to Baldur's Gate. There is no need to invite him into the group.

Then exit the region via the west road, towards Baldur's Gate. You're going to have some mandatory long rests at camp, with special events. After solving the relevant issues, you will finally arrive at your destination. Normally, Jaheira should have her own tent in your camp.

Chapter 3: Events Related to Jaheira

If you intend to develop a romance with Jaheira, it's best to keep her in your party permanently, so you have more opportunities to improve your relationship. Jaheira is full of humor, she appreciates jokes, but also behaving in a caring way. Helping refugees, townspeople, and doing good in general are all ways to gain his approval. You will have many opportunities to attend special events related to Jaheira, which are all opportunities to gain her favor:

- Pay a visit to the Harpers' base, hidden under the weapons store.

- Visit the Thieves Guild base in the Slums, to negotiate with 9 fingers. Then complete the entire series of quests linked to Minsc, detailed below.

- Visit Jaheira's adopted children in her house, northwest of the city. If Jaheira is not in your party, they will ask you to return with her. Remember to defend him during the dialogue that follows, and to use the Minstrel Badge to open the passage into his office. After deactivating the traps, you will need to find his secret hiding place and the rejuvenation ritual scroll.

- Going to the House of Sorrow in the northwest corner of the city with Shadowheart and Jaheira is a way to meet an old acquaintance, Viconia. If you spare it, Jaheira will appreciate it (but you'll lose some great items).

Sauver Minsk

The kind but simple-minded ranger also returns in Baldur's Gate 3, and Jaheira feels guilty after abandoning him during a desperate fight. Unfortunately, he is a victim of the Absolute, and of its mental control. If you want to be able to push things further with Jaheira, and incidentally, recruit Minsc, you will have to save him.

- By exploring the 9 Fingers office in the city's thieves' guild and reading the documents, you will learn that the Stone Lord, aka Minsc, is preparing to attack the Court of Auditors, aka the bank.

- Meet there. You have to convince one of the ticket agents to give you an access ticket. You can also steal one from the rooms in the stage. Using invisibility, or violence, to get past guards is also an option. When you reach the vault room, you will see Minsc escaping a Mimic, before fleeing the area into a portal.

- Once the fight against Bhaal's minions is won, search the area and question the corpse. Minsc escaped into the Sewer Tank, an area in the northwest corner of the map. All you have to do is go there, once again.

- You will find Minsc and the Absolute's minions in the tank, and the fight begins in any case. It is absolutely vital to use non-lethal damage. Open the "passive" tab of your characters' action bar, and activate the bubble. Non-lethal damage does not work with spells or ranged attacks, only melee. Instead of killing Minsc, this will knock him out when he drops to 0 health.

- When all the enemies are dead, a dialogue will begin, with Minsc becoming enraged. We must convince the Emperor to protect Minsc from the absolute. If Jaheira is present, she will half take care of it. When she threatens to destroy everything, respond, "I don't want to see if she's really going to do it."

- When Minsc wakes up, you have to show him that Jaheira is present and alive. If she is not in the group, you must send her visions of Jaheira via the larva. The other answers lead to mortal combat for the mindless ranger, lethal damage or not.

- You can then speak normally to Minsc, and help him find Boo, his hamster, in the next room.

- Back at the camp, you will be treated to a special scene during the night, between Jaheira and Minsc. Defend Jaheira's point of view as a priority, to obtain approval from him.

Jaheira's House

- You can go to the Jaheira house, in the northwest corner of the city. If Jaheira is not in your party, her adopted children will ask you to return with her. Consider defending him during the ensuing dialogue. Coordinates: X: -221, Y: -45.

- Climb upstairs and talk to the child in the room to get the Harp Pin.

- Use the Harp Pin. on the corner of the desk, in order to open the passage to the cellar.

- You can deactivate the traps by lifting the crate on the left, then interacting with the plate.

- Upon reaching the cabin at the bottom of the cave, Jaheira will tell you about her secret hiding place. Insist on exploring it.

- Open the wall cupboard, this reveals a room containing a very rare scimitar, and an amulet left by Khalid offering +1 Wisdom to Jaheira. There is also a strange rejuvenation ritual scroll placed on the furniture.

The Rejuvenation Ritual Scroll

You can immediately trigger a special dialogue by clicking on it, but the response options are limited. We have to wait until we return to camp to have a more serious discussion. Different options are then possible. You can encourage him to use it, in order to give him a second lease of life, or on the contrary, discourage him from doing so.

Despite our best efforts, we were unable to get her to use the rejuvenation ritual. The excuse is that our heroic deeds inspired her, and she doesn't feel she needs to escape death for the city to be protected. Ironically, our actions, intended to impress him, backfired in this context. We have not yet been able to confirm whether making other important decisions during the story, such as killing Tieflings, changes the outcome.

Is a romance possible with Jaheira?

We failed to unlock a romance with Jaheira. Even with a very high approval, and the completion of the various quests linked to it, the option never presented itself. The answer appears to be "No" at the moment. We also haven't found any trace of a romance online. The age difference may be the explanation. It is possible that special conditions must be met for this, such as successfully convincing her to use the rejuvenation ritual. More information on the subject may come in the future, RPG developers love to hide secrets of the genre.

GAYLE

Companions can be a bit of a pain at the start of Baldur's Gate 3, and even if he is friendly, Gayle will outright ask you to sacrifice magic items without explaining anything to you.

How to recruit Gayle?

- Unless you are specifically trying to avoid it, you are almost obliged to come across Gayle, while he is trapped in a Fast Travel Stone, a little after the wreck of the Nautiloid.

- You just have to choose to shoot his arm to save him.

- Be careful, if you play with Dark Urges, don't fantasize about cutting off its hand. At least, if you want to recruit him.

Combat Optimization

There's more than one way to play Gayle and optimize her. We advise you to offer him a respec, from Blight, to reallocate his starting characteristic points a little: 8/14/16/17/10/8 or 8/16/15/16/10/8 are good choices . The second option allows you to invest in the "Constitution Saving Throw Mastery" Gift to improve your concentration and obtain a round number on your Constitution.

- Wizard is a great class choice, and choosing the Evocation subclass is also ideal This allows you to throw Fireballs at the piles of enemies surrounding your group, without risking inflicting damage on your characters.

- Then all you have to do is invest in Intelligence for Gifts until you obtain 20 points, and favor evocation spells like Magic Missiles, Burning Rays, Fireball, Hail Flood, Wall of Fire , Cone of Frost, Chain of Lightning, and to devastate the opposing ranks.

- You can equip Gayle with light or even medium armor, as well as a Shield, to greatly improve her survival. Give him a staff with useful spells and spell bonuses as a weapon.

Romance with Gayle

This romance is very simple and straightforward compared to the others. Even players who have absolutely no intention of flirting with him risk going down this route by mistake.

- You must progress in Gayle's origin quest, which is obligatory if you wish to keep him in your group (see further).

- You need to have a good reputation with Gayle, which is very likely to happen, if you help her in her quest, and play someone nice, helping others.

- The first event linked to romance with Gayle takes place during chapter 1. He will offer to introduce you to magic. You can then imitate his incantations. If you are not interested, it is best to refuse to participate immediately.

- During Chapter 2, a magical image of Gayle will be found at camp, and he will invite you to stargaze with him as you follow the path. There is no real path to follow, this will automatically open the long rest screen, with the food reserves. By validating the thing, this will trigger a romantic scene with Gayle, during which he will confess his love to you. You can then formalize your relationship, or refuse.

How to improve your relationship with Gayle?

Gayle can be considered a good-aligned character (even if he is a bit pretentious). He approves of acts of kindness, of helping others and encouraging them. But what he really approves of is helping him. By agreeing to comply with his requests, you will easily rise in his esteem.

He hates cruelty, evil acts, or being left to deal with his problem.

Gayle's Origin Quest: Netheril's Orb of Destruction

❖ Chapter 1

After some time spent together (use the Long Rest function) or some relationship gains, Gayle will tell you about her "illness" which involves sucking magic from different magic items. It won't be long before he asks you for a magic item to sacrifice, which will be destroyed.

- You can recognize eligible items by their little red swirl icon, which mentions that it can be used to power Gayle.

- If you don't want to lose one of your current items, the best solution is to quickly go to the Druid Grove, and buy, or steal eligible magic items from the merchant, or Dammon.

- You have to agree to give Gayle a magic item, or persuade him to reach for it. In case of refusal or failure, he will leave the group.

- After giving Gayle 3 items, he will eventually calm down and stop begging for items. There's no need to panic, he's not going to do this the whole game. He will then tell you how he found himself in this situation.

- If Gayle dies, a magical image will appear, with instructions on how to bring him back to life. If you ignore the problem for 3 days, a cutscene will trigger at the camp, which will lead to a game over. But you really have to do it on purpose for it to happen.

❖ Chapter 2

- The beginning of chapter 2 is marked by the appearance of Elminster, the old wise man and chosen one of Mystra. It will help control Gayle's orb, while giving her the ability to cause its explosion. A new button will appear in the interface. You can activate it in combat, but this will cause an instant "Game Over".

- You have an opportunity to detonate the orb at the end of Chapter 2, when Ketheric is with the Chosen and the Mastermind. It is better to discourage him If it explodes, you will indeed trigger the end of the game and the end credits, but it's a bad ending.

❖ Chapter 3

- Gayle will mention the need to search for information on the Crown of Karsus. Go to the Lower City of Baldur's Gate, to the most remarkable shop: Magic & Sorceries.

- Talk to the librarian at the back of the store to obtain information on said book.

- The journey through the underground passages is quite complex; you must activate the magic levers in each section, teleporting to each wing of the area, and avoiding traps. After activating the Silver Hand and Elminster lever, you will have access to the wing of Karsus, and the precious book. This will trigger a dialogue, which you can direct in different ways: thirst for power, desire for redemption, etc.

- A few days later, Elminster will come and speak to Gayle at camp, asking her to go to the temple in the lower city to pray at the statue of Mystra, which will trigger an interview with the goddess. Once again, you are free to direct this, as you wish, depending on the desired conclusion.

Final battle and additional ending

At the very end of the game, when you reach the point allowing you to climb to the Infernal Brain, a special dialogue will be triggered with Gayle. You can ask him to climb up on his own, and blow himself up. This will bring the final battle to a quick and tragic end. Of course, Gayle is dead.

Epilogue

You are entitled to a final dialogue with Gayle, which can take different directions depending on different factors:

- If you have decided to give the Crown of Karsus to Raphael, things are looking bad for Gayle.

- Otherwise, he will try to recover it and repair it. But his motivations will depend on his relationship with Mystra. He may seek to give it to him to be treated, or to gain power and heal himself.

- In case of romance with Gayle, he envisages the future in your company.

Made in the USA
Monee, IL
08 November 2024